Building An Affordable House

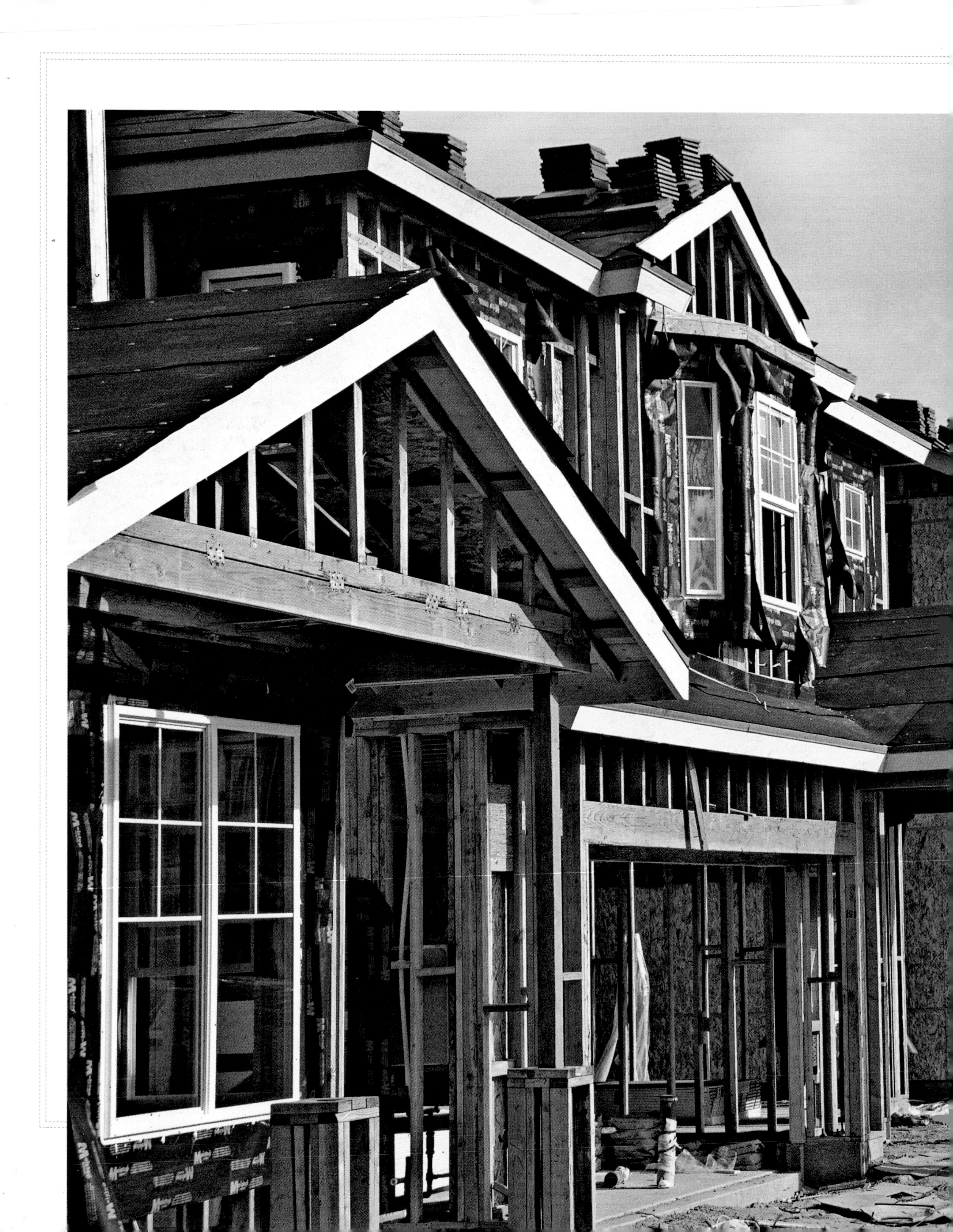

Building An Affordable House

Trade Secrets to High-Value, Low-Cost Construction

FERNANDO PAGÉS RUIZ

The Taunton Press

To my mother and father, Myrtha and Fernando Pagés, who taught me the value of hard work and the joy of learning. And to my wife Deborah and our sons Alexander and Fernando, who continue to provide the inspiration for everything I do.

The Taunton Press
Inspiration for hands-on living®

The Taunton Press, Inc., 63 South Main Street, PO Box 5506, Newtown, CT 06470-5506

e-mail: tp@taunton.com

Editor: Scott Gibson

Jacket/cover design: Michael Sund

Interior design: Laura Lind Design

Layout: Laura Lind Design

Illustrator: Chuck Lockhart

Photographer: Fernando Pagés Ruiz

Library of Congress Cataloging-in-Publication Data

Ruiz, Fernando Pagés.

Building an affordable house : trade secrets to high-value, low-cost construction / Fernando Pagés Ruiz.

p. cm.

ISBN 1-56158-596-3

1. House construction--Cost effectiveness. 2. Dwellings--Remodeling--Cost effectiveness. I. Title.

TH4815.8.R85 2005

690'.837--dc22

2004018121

Printed in the United States of America

10 9 8 7 6 5 4 3 2

Acknowledgments

It's to my employees and subcontractors that I owe the first and foremost word of recognition for enduring a year of interviews, photo sessions, and the unrelenting refrain, "Can't we do this for less?" In particular, I thank my employees Anselmo Mijangos, Johnny Valenzuela, and Walter Enriquez. I also thank Omar and Mario Lisak of Kasil Construction, Tom Trainor of Trainor Plumbing, Jerry Schidler and Robert Burns of Schidler's Electric, Dan Dalman of Dalman's Drywall, and George Roberts of Trademark Concrete Construction. These men are the heart and core of the best team I've ever worked with.

I regard my material suppliers as business partners and count on them to help me design and specify high-quality, cost-effective products. Without knowledgeable suppliers, modern housing would not exist, and neither would this book. I feel especially grateful to Jeff Taake, Mat Francine, and Ron Sipp of Millard Lumber; Dan Moser of Fisher-Moser Flooring; Ron Stump of Campbell's Kitchens; Paul Rohrs of Lincoln Winnelson; and Don Voges of Capital Concrete.

Although I strung the words, if this book were only about what I do it would encompass a very limited scope. The creators of this book include dozens of generous and knowledgeable builders, architects, and researchers who spent time advising and showing me examples of their work. In no particular order, I extend my sincerest appreciation to Donald MacDonald, FAIA, of MacDonald Architects, San Francisco; Don Hamill of Oakwood Homes, Denver; Gary Gafford of Beazer Homes USA, Southern California; Mark Kaufman, AIA, of Parkside Development, Houston; Kirk R. Malone of Mercedes Homes, Melbourne, Fla.; Bob Kayfus and Paul Sims of NuHome, Houston; Bill Eich of Bill Eich Construction, Spirit Lake, Iowa; Jim Schumacher of Woodinville Lumber, Woodinville, Wash.; and Dave Daniels of Houston Habitat for Humanity.

Without engineers, building scientists, and researchers, homebuilders might still be constructing thatched huts. In particular, I would like to thank Jay Crandall, P.E., of Applied Residential Engineering, West River, Md. Many of the techniques described in this book came from the minds and laboratories of the National Association of Home Builders' Research Center in Bowie, Md., and Steve Winter and Associates, Inc., in Norwalk, Conn. Both organizations provided me with generous advice and ample resources.

Of course, this book would not exist without The Taunton Press, which displayed reckless courage in accepting my proposal. I am especially indebted to my editor, Peter Chapman, who encouraged me to "write on" as he clipped, pasted, and cleaned up behind me.

Photo editor Wendi Mijal allowed me to take the pictures (which proved to be the most fun part of this project) and then, somehow, turned the amateur collection of snapshots I turned in into professional-looking photographs—I don't know how you did it, but thank you.

I would also like to thank my favorite *Fine Homebuilding* editor and friend, Tom O'Brien, for lobbying my book proposal, providing encouragement, and reading the original text.

Finally, I thank my friend Pamela S. Thompson for proofreading and providing invaluable editorial input and my photography teacher Barbara Hagen for showing me the light and teaching me to see the world through a lens.

Of course, there are many others who have contributed, and I would like to thank them each by name, but my editor says we've run out of space. Please know that I appreciate and acknowledge all of your help and advice.

Contents

Foreword

In 1951, shortly after I started my career building houses in southern California, my older brother, Jim, put $400 down and bought a brand new, three-bedroom, $7,000 house. His payments were $63 a month, including taxes and insurance. The American dream of owning a house was, maybe for the first time in history, possible for ordinary working people like us.

But prices rose and kept rising. For the average American, housing costs went from about 10% of annual income in the 1970s to 30% in the 80s. Nowadays, I've read, some people—strung out on credit cards and consumer loans—spend up to 60% of their income just for the roof over their heads. Such expenses may be bearable for a few, but most of the people I know (like my own children) have been priced out of the market.

So what's to be done? One way is the Habitat for Humanity model in which the owners acquire no-interest loans and help build their own homes. As a Habitat volunteer, I enjoy helping build these simple, decent, affordable houses here on the Oregon coast. The houses are basic but easy to heat, require little upkeep, and are affordable. That works for me. And thanks to this book, I've learned new ways that we can make our Habitat houses even more cost effective.

Building affordable is not a matter of cutting corners or reducing quality. It's not even a matter of working harder or faster. It is a matter of working smarter. And Fernando Ruiz's *Building An Affordable House* points us in that direction.

There are always new tools, new materials, and new methods to help builders compete. As builders in post-WWII California, my brothers and I were able to compete because we were able to adapt and change. Millions of people in our country need an affordable house. And that market is out there for enterprising builders who know they don't have to build monster homes to make a living. Ruiz's book is full of ways to accomplish that goal.

Larry Haun,
Fall 2004

How to Use This Book

If you want to know everything about using a high-value, low-cost approach to home building and remodeling, read this book from cover to cover. Or, if you prefer, skip around and read only the sections that apply to your specific project. Read the first two chapters thoroughly, because they provide an overview of the opportunities available for taking charge of construction expenses. The succeeding six chapters detail how it's done, trade by trade, and chapters nine and ten cover finishing details on the exterior and interior of your home.

The approach described here does not require sweat equity, purchasing schemes, or unproven technologies. It's a hard-dollar-cost, hard-nosed approach to saving money through knowledge. This is a book about empowerment that gives you the information you need to control every expense. If your home-building or remodeling contractor hasn't read this book, it's going to cost you thousands. One excellent way to save money is to have each trade read their corresponding chapter and then apply what they've learned to your home. I use every technique in this book, and I've never met a builder who could build better for less. Except, that is, until you.

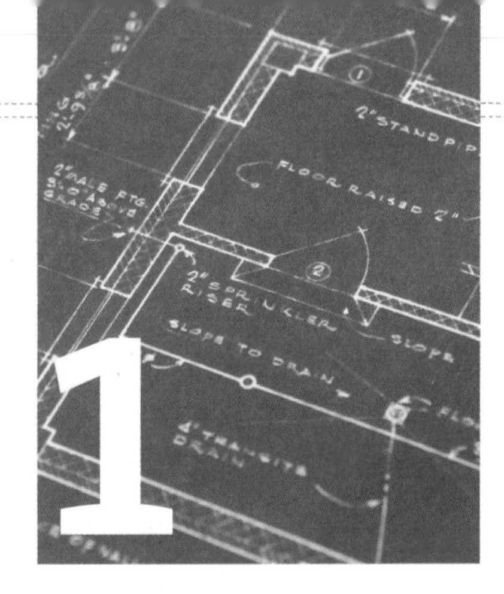

1 Money Matters

Since 1995, housing values across the United States have climbed by nearly 8.2% a year. In metropolitan areas, prices have swelled even more, sometimes at double or triple the national rate. A big part of this is due to spiraling construction costs. Sharp price spikes in lumber, labor shortages, and inflated subcontractor rates have driven the cost of building an average-quality home to nearly $100 per square foot (without land). Remodeling costs have climbed, too, making the alternative of adding on almost as prohibitive as moving up.

Under these circumstances, a builder who offers an economical alternative can develop a very successful market niche, one that ensures brisk sales in both good and bad economic times. This is why some of the largest and most successful builders in the United States, such

Driving past a neighborhood where families can afford to live tells the whole story: Good housing affords happiness.

What Is Value Engineering?

The catchall term value engineering describes a studied approach to cutting building costs without compromising quality. Value engineering entails comparing a range of construction methods and materials to determine the most cost-effective means to obtain a desired structural, aesthetic, or durability standard.

This tactic originated in manufacturing at General Motors® during World War II, when supply shortages forced GM to find substitute materials to produce certain military supplies. The stand-ins, such as steel alloys, often led to better products at reduced costs. After the war, GM pursued this tactic in its commercial lines. Other industries followed, and, by the early 1970s, the value-engineering approach finally came to home building.

as Kaufman & Broad Homes®, Lennar Corp., Centex Homes®, Mercedes Homes, and Beazer Homes USA focus their business on building the most competitive product in their market. As Patrick Hamill, CEO and president of Oakwood Homes® in Denver, told me, "We strive to offer our customers in every price range a lot more house for a lot less money."

It's hard to offer a precise definition of *affordability* because it runs the gamut from $85,000 starter homes in Utah to $850,000 "starter castles" in California. But in every market, there's an affordability niche to fill. The key to long-term success in constructing first-rate affordable homes of any price is learning to control costs without compromising quality—something more easily said than done.

In every market, there's an affordability niche to fill.

Controlling costs requires knowledge and skill. The affordable-home builder can't "just do it." It involves dissecting every facet of construction and searching for potential savings, then analyzing each dollar reduction against uncompromising standards.

A Hard-Dollar Cost Approach

As land development and regulatory fees have become ever-larger components of housing costs, not even the shrewdest builder can control every factor that boosts expenses. Hard-dollar costs remain one of the few variables where individual builders can actually make a difference. Many of the nation's largest homebuilders have implemented entire value-engineering departments to search for greater efficiency and lower costs. By challenging their suppliers and contractors to help cut expenses, Centex Homes managed to reduce its average cost of building a house by nearly 7%, about $10,000 on a median-priced house.

Many builders hesitate to incorporate state-of-the-art cost-control methods because they presume that entails compromising quality. They fear being labeled cheap, and with good reason. Indeed, if your method for building frugally amounts only to finding the lowest bidder and using the cheapest materials, then quality suffers. Cheap builders enjoy short careers.

The design and construction of a quality home on a reduced budget offers big challenges,

TRADE SECRETS

Developing the affordable "attitude" requires thinking like a grocer. With more than 1,000 line items in a typical construction budget, significant savings must be achieved one penny at a time. So, don't balk at a $10 reduction in overall costs any sooner than your corner grocer would give away 10¢ worth of produce.

although not insurmountable ones. Cost-conscious building becomes a lifestyle based on the Goldilocks principle: Neither too much nor too little; the balance has to remain "just right." The affordability equation explored in this book amounts to finding the best methods available to achieve extraordinary cost reductions while building a better house.

Even if you don't build "affordable" homes, you can still profit from learning the trade secrets of some of the most competitive home-builders in the industry. You can deploy powerful cost-cutting techniques to foundations, rough framing, and mechanical systems and then spend lavishly on granite countertops and gourmet appliances. Actually, some of the most competitive builders I interviewed while writing this book work at several price points simultaneously, and their affordable-home building techniques pay big dividends at the high end.

...if your method for building frugally amounts only to finding the lowest bidder and using the cheapest materials, then quality suffers.

The Affordability Equation

To build affordable homes, you have to start by figuring out what the term *affordable* means in your area. Affordable housing in Lincoln, Neb., ranges from $80,000 to $115,000. In San Francisco, it's from about $250,000 to $300,000. In either case, affordable housing by economic standards is housing that someone earning 80% of the median income can buy without exceeding 30% of his or her salary. For example, according to the U.S. Census Bureau, the median income for a family of four in New Mexico during 2003 was $47,314. This family can afford a home just under $150,000 with a monthly mortgage payment of $986 and taxes and insurance of $173. In this market, four-bedroom homes priced under $150,000 will sell quickly.

This formula becomes complicated when you weigh in taxes, insurance, utility costs, debt ratios, down payments, and everything else that influences the cost of homeownership. As a simple rule of thumb, check the classified ads in your local newspaper for the average rental rate of a three-bedroom apartment. You'll find this is close to what your local real-estate market defines as an affordable monthly payment. Use this monthly payment to work backward and figure out the optimal price tag for a three- or four-bedroom home in your area. Now turn to the classifieds and see how many new homes are selling at this rate. You'll find very few. But don't feel discouraged. Keep this optimal price in mind and make it your ultimate goal, even if at first it seems unreachable.

Penny Wisdom

The fundamentals of a balanced value-engineering approach include affordability, marketability, and durability.

AFFORDABILITY

- Start with a price target that fits your ideal market.
- Work backward to reverse-engineer an ideal budget based on this goal.
- Use value-engineering techniques to save on every facet of construction.

MARKETABILITY

- Strive to build the most house for the money.
- Make sure your design matches your customer's needs and wants.
- Put "value" where your customer can see it.

DURABILITY

- Simplify your design.
- Study and engineer every system in your home, then work out problems before breaking ground.
- Use advanced construction techniques to build a better shell.

TRADE SECRETS

Home-accounting software like Quicken® and Money® include fill-in-the-blank modules for calculating how big a mortgage you can afford at a given income and interest rate. Just plug in 80% of your city's median income (available from the U.S. Census Bureau at www.census.gov) along with the current interest rate for a 30-year loan, and *voilà*, the computer does the rest.

The Numbers Crunch

Housing values in the United States have grown from $8.4 trillion to $13.4 trillion since 1995, climbing at an average rate of 8.2% yearly. This represents a 45% to 52% increase in average value over a decade. It's little wonder, then, that Americans regard homeownership as a sure path to economic security and wealth.

Unfortunately, the relentless increases that helped buoy the financial status of existing homeowners also work to keep first-time buyers out of the market. Incomes have increased during the same period, but these increases trail housing prices. Low interest rates help offset this trend, but low interest rates won't last forever. If rates go up to 9% and beyond, many potential buyers will be priced out of the market.

Most mortgage companies believe that someone who spends more than 30% of his or her income on housing is financially vulnerable. Based on current lending parameters, a family would have to earn about $50,000 a year to buy a $125,000 home. At an interest rate of 9%, it would take $75,000 a year to achieve the same thing.

Although building affordable homes provides a cushion from interest rate swings, it also represents a social investment, providing the opportunity of homeownership to those whom the marketplace might otherwise leave behind. The intangible reward of knowing you are helping to build a better world goes toward building a better image for the entire home-building industry as well as your bank account.

TRADE SECRETS

By collaborating with your subcontractors and material suppliers, you can tap into an information-rich pool with some of the best cost consultants you can find anywhere at any price—and they'll help you for free. You'll also gain their ongoing personal investment in your vision.

Achieving Blueprint Democracy

When I decided to focus my home-building business on the affordability niche, I scoured the country, attending conferences and seminars in search of ideas. I visited affordable-home builders in several states. To my disappointment, most were relying on questionable accounting practices, subsidies, and tax breaks instead of building simple, economical houses. Almost by chance, I stumbled on a blue-collar approach to design developed by the Grand Rapids, Mich., Home Builders Association, which focused on practical construction methods instead of clever bookkeeping.

It all started when the Board of Realtors in Grand Rapids challenged the local home-builders association to develop an economical floor plan priced under $60,000. The builders grumbled, "It can't be done," but then promptly appointed a general contractor and one

BLUEPRINT DEMOCRACY

Having worked up a preliminary design for an affordable house, the author invites his subcontractors to find more savings.

TRADE SECRETS

Big builders save money by applying economies of scale to negotiate better prices. Small builders can form purchasing co-ops and do the same thing. For example, three builders can approach a garage-door vendor and offer to buy all of their overhead doors from him for one year in exchange for his best price and a $10 rebate on every unit. If one builder installs 20 doors a year, there's another $200 in his pocket.

representative from each subtrade to try anyway. Instead of hiring a designer to plot a house and then hoping that the bids would fit the budget, they went straight to the cost experts to come up with efficient, trade-by-trade designs from the start. I call their idea "blueprint democracy."

It can be difficult to get subcontractors to go along with value engineering and streamlined construction methods they're unfamiliar with. You have to sell the idea of affordable construction to your team. By engaging the trades early in the design process, you win their support. They become your consultants, providing ideas and buying into the success of your project.

> You have to sell the idea of affordable construction to your team. By engaging the trades early in the design process, you win their support.

Even the largest homebuilders engage their subcontractors in the early phases of design to streamline the construction process and make it trade-friendly. The basic value-engineering team consists of a designer, a framer, a plumber, an air-conditioning contractor, and an electrician. I have found it useful to include my lumberyard salesperson in the mix, too. A structural engineer and an energy consultant can come in later to help you refine the structural design, mechanical equipment specifications, and insulation package.

Every House Can Be Built Better

Even after you try to eliminate all the bugs up front, it's important to scrutinize the building process and purge your system of wasteful practices. Just like maintaining your weight,

Subcontractor Roundtable

To get my team of subcontractors hooked on affordability, I invited them to a "subcontractor roundtable" at a local steak house. I paid for lunch, and I told my subs that I wanted to build an affordable house that could be put up again and again. It would be a source of income in good times and bad. I asked this captive audience for help in designing every system for quality, efficiency, and reduced cost. Then I gave each subcontractor a preliminary floor plan on which to sketch his or her ideas. The mission was simple: Look for savings.

With more than 1,000 line items in a typical house budget, it's easy to find opportunities for small but cumulative cost reductions. However, I knew from experience that subcontractors are used to following plans, not creating them. They tend to view a builder's quest for cost reductions skeptically, since most of the time a builder's savings comes at the expense of a subcontractor's profit. I made sure my subs understood that I was looking for more efficient construction methods and not just lower bids.

To my surprise, my subs made suggestions that affected the entire design process, not just their own specialties. Many cost overruns come from inefficient overlaps between trades. For example, the air-conditioning contractor suggested changing the location of the utility room to a more central location, saving about $150 in ductwork. The electrician eliminated 6 in. of kitchen cabinet and moved a closet door 12 in. to eliminate two plugs, saving $110. A suggestion from the plumber shaved $300 off the budget.

TRADE SECRETS

Your job site is the proving ground for procedures and design. You must spend time observing on the job site to see if your value-engineering equations add up. Make changes when they don't.

implementing a frugal approach isn't something you do once and then forget. Following the initial effort, affordable-home building requires assuming a new lifestyle based on the conviction that every house you build can be built even better.

I once noticed two plumbers working on a waterline. After running a few errands, I returned to see what they had accomplished: They had installed a single sill cock. Two men, one hour, one sill cock. This equation didn't add up to savings, so I asked the plumbers why the job took so long. They had installed the sill cock where the plans indicated, which required drilling through 20 ft. of floor joists and soldering their way around a corner. We spent a few minutes looking for a better location for this sill cock and found we could plumb it in less than 30 minutes by moving it to a sidewall. I drove to my drafter's office immediately and had him change this fixture location on the plan in bold ink. This type of ongoing job-site revision can save hundreds of dollars in wasted materials and labor.

Sometimes this process translates into cost savings for the builder, sometimes only for the sub. I don't ask my subs to return every cent they save, even if the savings came from my suggestion. Instead, I ask them not to raise their prices. I try to get them to do what I do, build profits through added efficiency and not price hikes. When I help my subs achieve their financial goals, they are more likely to help me achieve mine.

When I help my subs achieve their financial goals, they are more likely to help me achieve mine.

Great Expectations

Once you have devised the most cost-effective home plan in the world, you're still stuck with trying to sell it to a consumer whose expectations are notoriously high. You can sell homes based on price alone, but unless you give buyers a house they can own with pride, you'll have to deal with their inevitable disappointment. This concern with value separates the affordable-home builder from the cheapskate. Remember, you're trying to build a home that your buyers will fall in love with and not just settle for.

With a front porch, some added glazing, and brighter colors, a simple house takes on curb appeal.

Meet your customer

Buyers looking for affordable homes come from all occupations and age levels, and increasingly from outside the United States. While their backgrounds differ, affordable-home buyers all want the advantages of homeownership without crushing mortgage payments. Otherwise, the specific housing requirements of each type of buyer vary. Young families and retirees represent the largest segment of the affordable-home market, but it's obvious that one size can't fit all buyers. Young families want plenty of floor space, a large yard,

Old-style affordable housing sometimes just looked plain and cheap.

and a two-car garage. Retirees might opt for less square footage, but they want proximity to shopping and medical facilities. Young families don't mind clambering up and down stairs; retirees want a single-story plan.

Minority ownership is another important factor. It began to surge in the 1990s, just when the baby bust would otherwise have dampened first-time homebuying activity. As a result, minorities accounted for more than 40% of net new owners from 1999 to 2004.

Before designing your affordable home, you have to research your market and decide whom you're building for. But within general parameters, you can guarantee the broadest appeal possible by staying within the boundaries of good taste—as defined by the community you're building for.

Before designing your affordable home, you have to research your market and decide whom you're building for.

Stick with plain vanilla

Homebuyers don't want to be the first on the block to buy a goofy-looking geodesic dome, even if it's cheap. Local market considerations should guide your initial design concept before value engineering molds the house to hit your per-square-foot dollar target. The neighborhood you're going to build in represents your best guide to the local market. Your home should fit in and look like it belongs there.

Most buyers and bankers prefer plain vanilla when it comes to housing. This conservative approach offends some designers and creative builders who are out to change the world. If you're one of these, have at it. You'll get over it after your first foreclosure. Your buyer's aesthetic needs come first, not yours. But if your market demands innovation, and some do, then provide it. In Los Angeles, I built inventive and whimsical neighborhoods that sold precisely because of their individuality. In Lincoln, Neb., a conservative town by any definition, I try to blend houses into the landscape inconspicuously.

Bare necessities are not enough

Not long ago, affordable housing meant a cracker box with thin walls, no insulation, one bath, and a radiant wall heater. But times have changed. Even the most unassuming homebuyers expect some curb appeal, an efficient central air-conditioning system, brand-name appliances, and amenities like decks, landscaping, and an automatic garage-door opener. Plan for the most common luxury items, even if you offer these items as upgrades or improvements that the buyer can install later.

I always include wiring for an automatic garage-door opener, junction boxes approved for ceiling fans in each bedroom, a ledger for a future deck, and plumbing for a future sprinkler system. Sometimes I frame a portion of the house for future bedrooms and family areas and rough in the plumbing for an extra bath. These thoughtful touches add cost, but they become necessary when trying to sell homes on the open market. Though a buyer's present circumstances may limit his or her means, we live in a land of hope and prosperity. When you provide options for improvement, your buyer can reconcile forgoing a few luxuries temporarily.

> By and large, immigrant and minority buyers are looking for the same thing every homebuyer wants: value.

Don't overlook infill sites

Before you can build an affordable home of any description, you have to locate an inexpensive plot of land. Young families will move almost anywhere they can find safe streets, a good school, and ready access to work. For this reason, I believe it's easier for the small builder to forgo the retirement crowd and focus on the first-time buyer's market, which allows greatest flexibility in the choice of lots.

Scattered lots in existing neighborhoods, dubbed "infill sites" in real-estate jargon, represent a prime source of inexpensive land. You can find infill sites in all kinds of neighborhoods, but most often in ethnic neighborhoods that many builders avoid for fear of investing in distressed areas. My experience contradicts this view. According to a report by the Joint Center for Housing Studies of Harvard University, the hottest groundswell of first-time homebuyers comes from immigrant and minority populations.

In areas with distressed real-estate markets, builders who participate in the creation of high-quality affordable housing can get substantial assistance from government sources, including free land and favorable financing.

Though this book describes a bricks-and-mortar approach to building economical homes, the combination of municipal subsidy with intelligent building practices can benefit both the neighborhood and the homebuilder.

The bottom line is value

By and large, immigrant and minority buyers are looking for the same thing every homebuyer wants: value. But aesthetic standards vary between cultures, and if you do a little research, you can often find pockets of opportunity ignored by others. In Los Angeles, I built affordable homes right on a busy major thoroughfare. Conventional wisdom says buyers don't want to live on busy streets, but many people from other cultures may feel right at home in this environment.

These experiences taught me that traditional real-estate rules don't always apply. The experts in home design are your buyers.

New-style affordable housing looks affluent and tasteful.

2 Design

If you've ever worked with architects, you know they like to talk about things like "design grammar" and "architectural rhythm." These fancy-sounding terms refer to the aesthetic guidelines architects use to create a cohesive blueprint. Because of the principles defining each architectural style, you would never confuse a Frank Lloyd Wright building with one designed by Frank Gehry. Each architect draws from a different "kit" of design principles. Likewise, a kit of belt-tightening design principles can help you nail down a cost-effective blueprint no matter what style of architecture you choose.

Some developers call this affordability-by-design approach the "front-loaded" method. Instead of working backward to save money on a desirable floor plan, you start by sketching the very first draft of your plan using a palette of cost-cutting options that virtually ensures

Anderson and Ora Lee Harris's 600-sq.-ft. "butterfly house" was designed and built by Samuel Mockbee's Rural Studio in Mason's Bend, Miss., for $25,000. The intersecting rectangles that make this structure appear poised for flight cover a large screened porch and collect rainwater for use in toilets and laundry.

Modern architecture, with its clean lines and austere aesthetic, lends itself to affordable design. This townhouse, designed and built by Donald MacDonald, sold for about $120,000 in the mid-1990s, a remarkably low price for San Francisco.

This spacious, whimsical interior, designed by architects Cathi and Steven House, accents a sophisticated plan with a sensible heart. Built for less than $77 per sq. ft. in San Francisco's Nob Hill neighborhood, the house features a square shape, which the designers chose to keep costs low, an affordable élan with a curved wall, carefully chosen accent colors, and a little breathing-room height in the living-room ceiling.

efficiency. Although this method makes economy the primary goal, it doesn't preclude architectural creativity. The homes illustrated in this chapter prove that you can be innovative while saving money. But when combining flair with frugality, you have to understand the architectural grammar of affordability and live within its restrictions. Economy is built into a project from the foundation up, not added as an afterthought. Affordable architecture requires weighing many options and consumes a lot more time than architecture as usual.

Economy is built into a project from the foundation up, not added as an afterthought.

Not every architect is up to this task. Many architects view affordability as a hindrance to creativity. Only a few see the challenge as a framework in which their ideas can flourish. The purpose of setting standards in affordable architecture is to facilitate frugality, not boredom. Once you know the basics of cost-conscious design, you can bend the rules and add pizzazz where it counts, creating maximum aesthetic effect at minimum cost. But before you try designing outside of the box, learn to live within it. You'll find a nicely appointed box isn't a bad place to start.

Building a Better Box

A cube yields the highest floor-to-wall ratio. In other words, it provides the smallest area of exterior surface to cover a given living space. By limiting the total area of exterior wall, the box eliminates the additional foundation, framing, insulation, siding, and other components that make exterior walls an expensive element. That's why I always begin by designing the exterior shell, not the floor plan.

More Can Be Less

We often think that cost control means smaller homes. To an extent, this is true. But simple shapes usually mean lower costs even as sizes increase. In order to cram a kitchen, dining area, living room, stairway, and entry into a tight, low-square-footage floor plan, you often end up with a complex shape. This complicates the second story and usually requires many intersecting rooflines and jogs that push costs up.

By opening up the structure and simplifying the shell, a slightly larger home can actually cost less to build. Don't hesitate to allow floor space to increase slightly to achieve simplicity. Sometimes a few extra square feet can actually bring the overall price down.

TRADE SECRETS

In hurricane and seismic zones, a basic box represents the most stable structural shape, reducing the need for expensive reinforcement. Symmetrical boxlike buildings with continuous floors and small window and door openings have high lateral and shear resistance.

Because interest in affordable housing peaked right after World War II and then began to wane again during the 1960s, more examples of affordable design exist in older neighborhoods than in new ones.

Ratio of Floor Area to Wall Length for Rectangular-Shaped Floor Plans with an Area of Approximately 1,100 Sq. Ft.

Plan C and Plan E represent highly efficient shell solutions, but the narrow floor depth of Plan C makes it preferable to the wider Plan E because shorter joist and rafter spans are more economical to build.

Plan	Size	Floor area	Exterior wall	Floor-to-wall ratio
A	24 ft. by 46 ft.	1,104 sq. ft.	140 lin. ft.	7.89
B	26 ft. by 42 ft.	1,092 sq. ft.	136 lin. ft.	8.03
C	28 ft. by 40 ft.	1,120 sq. ft.	136 lin. ft.	8.24
D	30 ft. by 36 ft.	1,080 sq. ft.	132 lin. ft.	8.18
E	32 ft. by 34 ft.	1,088 sq. ft.	132 lin. ft.	8.24

Adapted from *Reducing Home Building Costs with OVE Design and Construction*, NAHB Research Center, Inc. (Upper Marlboro, Md., 1977).

Designing an economical shell

The shell of a home refers to the three surfaces that separate the indoors from outdoors—foundation, roof, and exterior walls. These components determine the basic cost of a home more directly than the size and distribution of interior rooms. Of the shell's three components, the roof and foundation are most expensive. So the more living area you contain within a given footprint, the lower the cost. That's why a two-story house costs less to build than a one-story of the same square footage.

The equation gets a little more complicated when you factor in the exterior walls. Certain rectangles work almost as well as squares. And when you take into account the practical limitations of floor and roof spans, rectangles sometimes work much better. For the sake of building an economical frame, it's always better to keep the breadth of the structure within standard floor and roof spans that allow the use of off-the-shelf framing materials. So beyond a 32-ft. square, a rectangle becomes the best option. When the house must get bigger, add width perpendicular to the framing instead of depth.

Complex footprints like H- and C-shapes never represent a good option from the affordability perspective. These arrangements have more corners, complex roofs, and much worse floor-to-wall ratios than any four-sided structure. For more information on the efficiency of shapes, see Appendix 1 on p. 196.

Built in Denver by Edward B. Hawkins and Eugene Sternberg, this house represents one of many inspired by Frank Lloyd Wright's Usonian concept of democratic architecture. Between 1949 and 1957, Hawkins developed an entire neighborhood, called Arapahoe Acres, using the Usonian model.

WHY SQUARE SHAPES WORK

The square box contains 625 sq. ft. of floor space within 100 ft. of exterior wall, while the rectangular box encloses the same 625 sq. ft. but requires 125 ft. of wall. The rectangular box costs approximately 25% more to build.

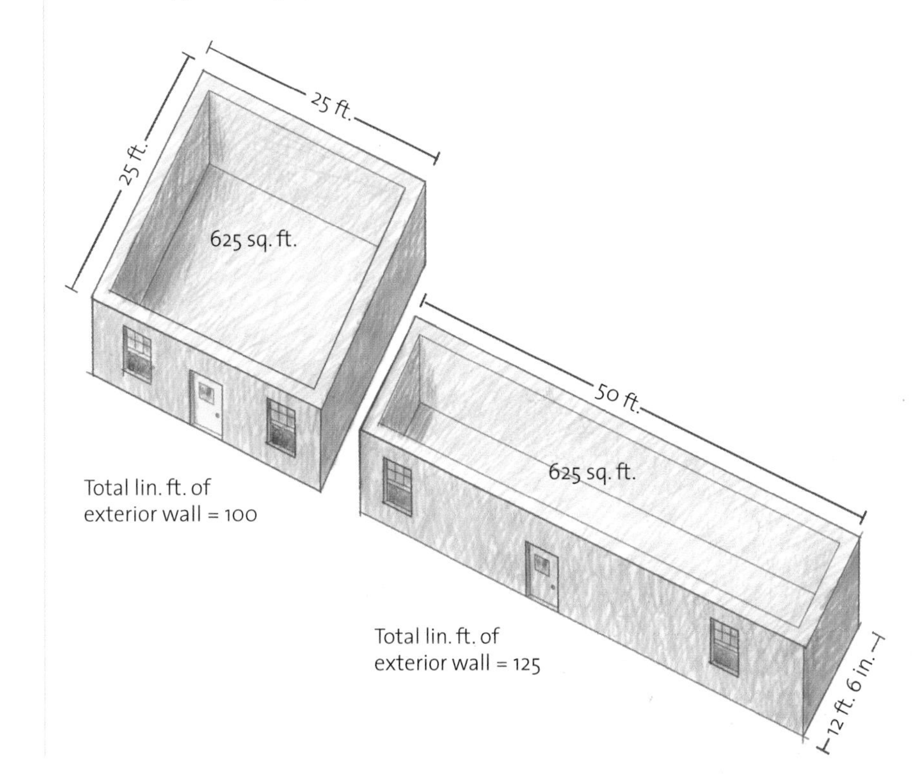

Using a Modular Planning Grid

Unlike many of his peers, Frank Lloyd Wright saw the creation of simple, affordable homes as a noble mission. He drew scores of compact, cleverly designed suburban houses that he eventually christened *Usonian* to describe their unique "United Statesian" character. From carports to slab-on-grade construction, Wright invented much of what we take for granted in production housing. Sears & Roebuck incorporated many of Wright's ideas into inexpensive precut housing packages from 1916 to 1933, and then tract housing developments throughout the country adopted them.

To optimize the use of standard building materials, Wright pioneered the concept of modular design. He drew blueprints on a horizontal 2-ft. by 4-ft. matrix that helped him corral his design ideas within the practical limits of inexpensive products like plywood and concrete. This modular approach also helped him take advantage of off-site and factory-built components, further reducing the cost of construction.

Many innovative concepts, such as modular design, slab-on-grade foundations, carports, spacious "California" interiors, corner windows, and cantilever rooflines began with Frank Lloyd Wright and later made their way into production housing.

TRADE SECRETS

I print my plans at ¼-in. scale on standard 18-in. by 24-in. sheets for easy reading and handling on the job site. I also print copies at ⅛-in. scale on 11-in. by 17-in. sheets. The smaller size fits into my office files as well as a conventional envelope when I need to mail plans to customers and vendors.

Thanks to modular design, this plan works as a single-family home or duplex without any change to the structure except the addition of a one-hour firewall between units.

Since you pour a lot of cash into the foundation, it pays to spend a lot of time fine-tuning the design.

Building materials, such as studs and plywood, still come in multiples of 4 ft. to accommodate traditional 16-in. on-center framing. But more and more building components, like trusses, lay out in 24-in. units. Since 4-ft. and 2-ft. dimensions work interchangeably, the planning grid that accommodates the widest range of building materials from foundation through finish is a flexible 2-ft. by 2-ft. module. This is especially true when you take advantage of 24-in. on-center Optimum Value Engineering (OVE) framing, one of the most profitable construction systems available to the modern builder (see Chapter 4).

Even when working with a 2-ft. by 2-ft. module, make sure the depth of the house (the dimension perpendicular to the joists) remains a multiple of 4 ft. (such as 12, 16, 20, 24, or 32). This maximizes conventional structural joist spans and 4-ft. sheathing widths and helps reduce waste.

Likewise, calculate rafter lengths to accommodate full widths of 4-ft. sheathing, and design the span of the roof in a 96-in. module that contains the material lengthwise. Sometimes, just changing the roof overhang by a few inches or playing with the roof pitch (who says you can't use 3¾ in 12 instead of 4 in 12?) can mean the difference between scraps of sheathing destined for the garbage bin and optimal material management.

To maximize the payback of the grid system, every window, door, stairway, cabinet, wall intersection, and other building component must be designed on the same module. Every time you deviate from the modular pattern, you waste materials and money. This becomes critical when dealing with structural components, such as framing. As long as the floor, wall, and roof members fall on the same modular layout, loads are transferred efficiently from roof to footings. When stairways, windows, and other openings are placed on module, you interrupt the fewest framing members and eliminate the need for structural redundancies and big headers.

Besides providing an economical layout from frame through finish, modular design allows you to reconfigure the building blocks of your home to vary the shape and make use of different lot patterns. For example, if you view your house and garage as two independent design blocks, you can usually create two or more versions of the same plan to accommodate different lots by moving the blocks around.

Builders using this approach will find real advantages. Interior spaces can be reconfigured to accommodate different floor plans for different buyers. For example, I have one plan that easily converts from narrow (24 ft. by 32 ft.) to wide (32 ft. by 24 ft.); includes two, three, or four bedrooms; and works equally well as a single-family home, duplex, or modular townhouse complex without appreciably changing the design or unit cost of construction.

Keep the footprint simple

Buyers don't value elaborate foundations as much as they appreciate soaring rooflines, large windows, or expensive details. So keep your footprint as small and simple as possible. You'll find it's less expensive to cantilever a wood-framed floor to add square footage than it is to enlarge the basement. For example, if you're building a 36-ft.-wide by 24-ft.-deep house, place it on a 36-ft. by 20-ft. foundation and cantilever 2 ft. front and back. This saves 45 cu. ft. of digging, 8 ft. of footing and foundation wall, 64 sq. ft. of waterproofing, and 144 sq. ft. of basement slab. You can usually add the cantilever discreetly at the rear of the house, especially if you have a deck to hide the offset, or use a cantilever to accommodate closets, bay windows, nooks, and fireplaces that don't require support directly underneath.

It's less expensive to cantilever a wood-framed floor to add square footage than it is to enlarge the basement.

Cantilevered floor framing allows additional square footage without incurring additional foundation costs.

DESIGNING ON A GRID

Frank Lloyd Wright used a 2-ft. by 4-ft. modular grid to lay out his floor plans and elevations, marking off the squares directly on the concrete slab to make construction easier. The plans sometimes did not include dimensions. The grid provided builders with the information they needed to erect the structure.

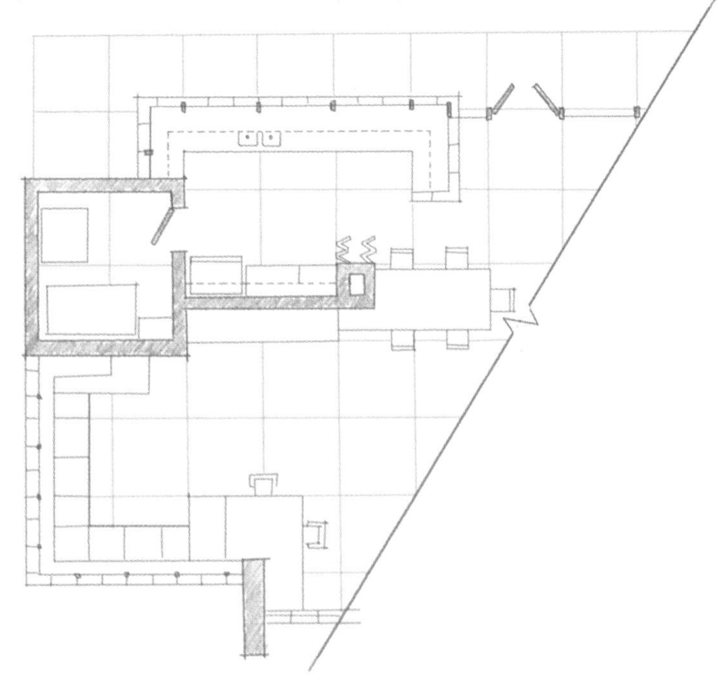

TRADE SECRETS

When possible, design the foundation with a beam or bearing wall down the center to split the depth of the house and reduce the load on your joists. A center-bearing wall distributes floor loads more efficiently because it actually doubles the capacity of exterior walls by balancing half the total weight on a single central-bearing structure.

A simple square shape becomes charming with a few attractive details, such as this decorative awning.

TRADE SECRETS

From an interior view, a bright window in the center of a room makes the adjacent walls look dark and the entire space feel small. Place the same window off-center, toward one corner, and the room appears bigger as light bounces off and washes along the adjacent wall.

Carpenters install panelized foundation sections made from treated wood for a house that will have a crawl space instead of a full basement. Foundations should be designed to carry design loads, but overbuilding is a waste of money.

Don't overbuild the foundation

Foundations can vary considerably, from slab-on-grade and shallow frost-protected designs to crawl spaces made from foundation-grade wood and full basements formed with reinforced concrete. (These options are discussed thoroughly in Chapter 3.)

Whatever system you ultimately choose, make sure the foundation performs the function required—namely, holding up the house—without being overbuilt. Since you pour plenty of hard cash into the foundation, it pays to spend a lot of time fine-tuning the design. For the price of an average overbuilt basement, you could add two bedrooms and a bath. Yet an unfinished basement adds nothing to the value of your home, while each bedroom boosts your equity by $10,000 or more.

An opening of 6 ft. or less on a gable-end wall may not require a structural header at all.

On exterior walls, follow the grid

Exterior walls, with their costly insulation, sheathing, and siding, make up the third most expensive component in the structural shell. Once you've maximized the ratio of interior floor space to exterior wall space, you can economize even further by simplifying each elevation.

Whenever possible, locate large openings, such as garage entries and patio doors, in nonbearing gable ends to minimize header and wall-framing requirements. Gable-end walls, when used in conjunction with parallel trusses, carry only a small portion of the roof load. In most areas, an opening of 6 ft. or less on a gable-end wall may not require a structural header at all. Along bearing walls that run perpendicular to roof framing, use standard windows of 4 ft. or less in width and make sure to stack at least one side of the opening on stud layout from floor-to-floor, even if windows end up slightly off center in interior rooms.

Even if you have to move a window several feet to stay on stud layout, do it. In a small bedroom, this placement might actually provide more space for furniture. Builders often design their houses without consideration for furnishings, which results in rooms that look great empty, but become impractical to live in.

By using a "kit of parts" approach, builders can offer customers easy upgrades without having to redesign every time they build.

Use exterior upgrades sparingly

No matter how affordable your house is, it will be difficult to sell if it doesn't have an attractive facade. But curb appeal generally refers to the front elevation (or front and side on a corner lot). Here, you may opt to splurge on larger pop-out windows and elevation changes regardless of stud layout or other cost considerations. But first, design the simplest and most cost-effective elevation you can. Only after you've controlled costs can you add design elements like brick, window reveals, shutters, shingles, and subtle elevation changes in much the same way you would snap noses and ears on Mr. Potato Head®.

By using a "kit of parts" approach, builders can offer customers easy upgrades without having to redesign every time they build. It also allows builders to construct the same house several times in a neighborhood without seeming redundant.

Get rid of unnecessary windows

Windows cost about twice as much per square foot as an exterior wall, so eliminating unnecessary windows can save hundreds of dollars,

At Rockhill Gardens in Kansas City, Mo., one developer built an entire neighborhood using one floor plan and clever elevation changes. The result is a varied and lively streetscape.

The Overbuilt Foundation

Foundations tend to be one of the most overengineered areas of a home because foundation codes were designed assuming worst-case soil conditions. In many cases, foundation systems exceed actual requirements by a factor of 10 or more. While a typical two-story home with a full basement transfers a load of 500 lb. to 1,500 lb. per linear foot, a typical 8-in.-thick basement wall on a 12-in. by 16-in. footing can handle about 80,000 lb. per linear foot.

TRADE SECRETS

Elegance involves effective understatement, not pretension. For example, change the lap width of your siding to highlight architectural features such as gable ends, chimneys, and projections, achieving subtle textural variety and detailing that conveys sophistication without extra expense.

TRADE SECRETS

When selecting more elaborate roof trusses, make sure the truss manufacturer can provide upgrades that don't burden your carpenters. Site-framed fillers and filigree, such as hips and fancy soffits, take a lot of extra time to build.

Long diagonal views that allow the eye to travel outdoors create a sense of inner space even within the confines of a small house.

By placing a large pane of glass at the entry to this tiny bedroom, architect Donald MacDonald creates breathing space in cramped quarters.

even on a small house. Windows are also less energy efficient than walls, which means that reducing glazing reduces heating and air-conditioning requirements, too. In seismic and hurricane zones, eliminating windows, especially on the first floor of a two-story house, adds stability to the structure and reduces the need for reinforcement. On small, narrow lots, a windowless side-yard elevation adds privacy. Careful window placement designed to create broad diagonal views inside rooms and to maximize exterior vistas can actually make a house feel more spacious and airy than wall-to-wall glass.

Rooflines add drama

By varying rooflines, you can change the appearance of a house even more dramatically than by changing wall elevations. Distinctive roofs add a touch of luxury to otherwise simple structures. And although this comes at a steep cost, it's one your customers can see and appreciate.

Roof trusses have made more expensive roof designs, such as gambrels and hips, affordable even on economy homes. But on second stories, or wherever rooflines are not prominent, it's a good idea to use a simple, low-pitched roof not exceeding a 5/12 pitch. Add interest with snap-on mansard hips, faux gambrels, and gable ends. Using decorative details to break up large spans of roof gives a house a more comfortable, cottage-like scale.

At small, intersecting gables over windows and architectural features, you can splurge on extravagant, steep pitches because you can build these miniroofs without too much extra expense. However, try to avoid true intersecting gables, such as those usually constructed over L-shaped homes, because these sharply increase the cost of framing.

Other inexpensive variations include the half-hip, Dutch gable, and shed roof framed over a front entry, pop-out window, and framed projection. Extended eaves over

A roof-to-wall junction like this one costs less to build than an intersecting gable roof.

TRADE SECRETS

Attic trusses are more useful over the garage of a two-story house, where you can fit a door into the attic from a second-floor room or hallway without building a stairway.

Although these twin townhouses have identical floor plans, different rooflines make them appear distinct.

porches and entries work well, too. But full-hip roofs and open-rafter designs like the gambrel are not cost-effective.

Attic trusses represent another useful but pricey alternative. Homebuyers like attic areas for storage. Some developers even install a stairway to the attic so homeowners can use the extra space comfortably or finish the space in the future. This, however, is one of the most expensive ways of achieving flex space (see "Handling Upgrades and Customization" on p. 23).

Inner Space

Once you have completed a first-draft sketch of the house, a room-by-room analysis of its interior will identify its strengths and weaknesses. Just as buyers want curb appeal, they also want a house that feels spacious and comfortable inside. In a small house, this means combining rooms into large living environments. By eliminating interior walls, you cut down on framing, electrical, drywall, trim, painting, and even air-conditioning costs.

Steep roof pitches, enclosed soffits, Greek returns, valleys, and complex intersections added about 10% to the cost of building this house.

Simple elements such as gables, window pop-outs, and a front porch give this otherwise rectangular home by William Lyon Homes, Inc. of Huntington Beach, Calif., a rich textural appeal.

TRADE SECRETS

Consider the dimensions of your finished floor material when you design room sizes. Carpet comes in 12-ft. rolls. If your room measures 14 sq. ft., your carpet installer has to add a 2-ft. seam down one end of the room.

As with the exterior shell, the relationship between the floor area and the total linear footage of interior walls tells you a lot about a home's cost efficiency. A higher floor-to-interior-wall ratio represents a more efficient design. But as you reduce the overall size of a house, you still should provide utility within each square foot of space. For instance, if the house is designed with long diagonal views, well-placed windows, and no extraneous walls, it will feel spacious and airy, even without a vestibule tall enough to accommodate a pet giraffe.

Keep living spaces open

When you blend a number of rooms—say kitchen, dining room, and family room—into a combined area, don't forget to add a touch of architectural drama to highlight its function as a gathering space. You can achieve this with a larger window, a change in ceiling height, or a border of contrasting floor finish, such as a hardwood perimeter.

Whether the house is large or small, the perception of space ends at the wall. Try not to block sight lines. When you can look through wide-open interior spaces to the great outdoors, you feel less restricted.

In areas prone to flooding, the first floor may have to be built above the 100-year flood mark, increasing the height of the house and making a second story look awkward. Attic trusses allow a lower profile.

Once the basic floor plan is complete, it's time for a careful room-by-room assessment that will uncover the strengths and weaknesses of the design.

TRADE SECRETS

Vinyl comes in 12-ft.- and 6-ft.-wide rolls, so it affords a little more flexibility than carpet. But it's still important to keep sheet-good dimensions in mind. For example, an 8-ft. by 6-ft. bathroom will require an 8-ft. piece of 6-ft.-wide vinyl—or 5.33 sq. yd. But if you increase the bathroom's width to 7 ft., you'll need 10 sq. yd. of vinyl, almost doubling the material required just to accommodate an additional 8 sq. ft. of floor area. Bad choice.

BUILDER'S CORNER

Handling Upgrades and Customization

To lure buyers, builders often include luxury features that make their basic models more attractive than the competition's. Developers call this tactic "market creep." A major part of saving money in construction is leaving out luxury items such as fireplaces, sun decks, and even garbage disposals and automatic garage-door openers. When upgrades become standard equipment, buyer expectations rise along with the price.

If you're a builder, offer buyers features they really want by taking an "options" approach to upgrades. Construct a bare-bones house and offer a menu of features that buyers can choose from and pay for. After all, no one really gives away free upgrades, they just add them stealthily to the base price. Builders can get a good sense of what people want by trying to find out why a house hasn't sold. Ask potential buyers what's missing. Soon, you'll have a comprehensive wish list to work from.

Make upgrades easy by planning every option with the same design methods you used in drafting the original blueprint. Talk to subcontractors and suppliers and figure out how to make changes without throwing off the production schedule and base price. Draw the changes into the floor plan as dotted-line options and see how they integrate with the basic structure. Note any rough-in required to install these upgrades after construction and include the most common ones—like prewiring a disposal outlet under the kitchen sink—in your basic plan.

Try to find easy ways to customize finishes even when you've nearly completed construction. For example, I paint all eaves and trim white. When a buyer chooses a two-tone color upgrade, all I need to do is brush the fascia, shutters, and front door. The whole process takes about four hours. I charge the buyer a reasonable price and make a little extra money for the trouble.

Likewise, I install code-compliant metal junction boxes in each bedroom ceiling for an easy paddle fan upgrade. I rough in the wiring for an automatic garage-door opener and lay out the family-room framing to accommodate a fireplace option. By expecting change orders and even offering them as another aspect of the "kit of parts" approach to embellishing your basic box, you turn this bane of every builder into a net benefit for all.

When builders begin adding features to lure more customers, prices are sure to rise. Builders call the phenomenon "market creep."

By using an open tread design, this stair preserves light and an interior view of the outdoors in a narrow living room.

TRADE SECRETS

Checking the relationship between the floor area and the linear footage of interior wall space reveals a home's cost efficiency. My 816-sq.-ft. home has 75 ft. of interior wall with a floor-to-wall ratio of 11 (816 divided by 75 = 10.88). If I added a wall around the kitchen, my floor-to-wall ratio would drop to 9%—a 20% increase in relative cost.

Give children adequate bedrooms

Despite our tendency to showcase adult bedrooms, it's children who need more space to play. I have found that families prefer a master bedroom big enough for a queen-sized bed, along with ample secondary bedrooms for the kids. You can maximize bedroom area by eliminating any wasted space, such as those short entry corridors commonly found between bedroom doors and an adjacent closet. Try to build closets as bump-outs, rather than insets that take up valuable living area. Cantilever the closet if necessary. Take advantage of found spaces under stairs or in attic areas to create additional storage. Whenever possible, locate windows near the corner, facing an unobstructed view.

Despite our tendency to showcase adult bedrooms, it's children who need more space to play.

TRADE SECRETS

Few upgrades deliver as much benefit at low cost as airtight construction and upgraded insulation. Many times, you can offset the added expense of a weather-sealed shell with substantial savings in HVAC sizing requirements.

Hallways are wasted space

Though you need to move from room to room, hallways and circulation areas should be considered expensive, wasted space. A 2,000-sq.-ft. house with 400 sq. ft. in halls is actually smaller than a 1,800-sq.-ft. design with 100 sq. ft. of hallways. Calculate the percentage of your house devoted to passageways by highlighting hallways, stairwells, and circulation areas on your floor plan. It should not exceed 6% of the finished space. Anything above 9% represents unnecessary waste.

Stairways should be straight runs

Stairways are one of the more expensive components in a home. In general, try to design straight-run stairs without separate landings. Frame stairs parallel to floor framing and on module to avoid interrupting joist layout.

Careful planning can save up to three joists. For example, doubling the framing on both sides of the opening is not necessary when the stair header is located within 4 ft. of the end of the floor span. A single header is usually adequate for openings up to 4 ft. wide. (For more details, see pp. 81–83 in Chapter 4.)

Bathrooms can't be too cramped

It's important to make bathrooms appear large and comfortable enough for two people to use simultaneously. You're better off with two comfortable bathrooms than three small ones. Most buyers will happily forgo a powder room for a full-sized bathroom shared by two bedrooms. When possible, provide privacy with a separate water closet for the toilet. Make sure vanities have plenty of counter and storage space.

You're better off with two comfortable bathrooms than three small ones.

You can make a bathroom feel bigger by providing a full-sized mirror over the cabinet. Install a good ventilation fan and, whenever possible, a window. A skylight can make a small, ordinary bathroom feel spacious. Factory-made skylights have become inexpensive enough to consider even in economy houses.

Make the kitchen bright

Although gourmet kitchens are in vogue, this room doesn't have to be glamorous or full of gadgets, just practical and bright. Most people prefer light woods, long countertops, and an island with room for bar stools.

If you build the kitchen into the open family space, it will appear large and airy even when it's small. Careful lighting and window placement add to the feeling of breathing space and comfort. When the design doesn't call for many cabinets, consider a built-in pantry. This item alone often sells a house.

Try to lay out your cabinet and countertop runs in 3-in. increments to take advantage of standard factory lines. Simple door styles without knobs and preformed countertops look clean and modern. Offer features like lazy Susans, appliance garages, and cutting boards as upgrades. Most buyers consider a dishwasher and self-cleaning oven standard equipment, but you can skip the microwave range hood and garbage disposal.

Laundry doubles as mechanical room

In a small house, the laundry room can provide an unfinished space for mechanical equipment and long-term storage. It should be located in the plumbing cluster, near the kitchen and baths. In a slab-on-grade design, place the laundry in a closet off the kitchen or next to a hall bath. A stacked washer and dryer saves space, but these appliances can cost a third more than side-by-side models, and they do not vent or plumb as easily.

TRADE SECRETS

Shower stalls and three-quarter baths are often proposed as inexpensive alternatives to full baths—but that's bad advice. The only reasons to install a shower in lieu of a full tub-and-shower combination are space limitations and buyer preference. The additional framing and glazed shower enclosure add substantial cost over a standard 5-ft. tub-and-shower combination with a curtain rod.

Educating the Architect

I once met with clients who wanted to build an architecturally unique home at a reduced cost. Their architect had sketched a few ideas on paper, and they were very excited with the results. I looked these sketches over and then delivered the bad news: The designs were beautiful, but they were inherently expensive.

Although you can achieve savings on any floor plan by using value-engineering techniques and selecting economical materials, it's the basic blueprint, more than anything else, that determines the total cost. Once my clients' architect incorporated my design methods described in this chapter, his revised drawings delighted my clients, especially when I told them their $250,000 home would only cost $180,000 to build.

Don't be afraid of offering a one-car garage. Oakwood Homes in Denver, as well as other builders, sell single-stall garages on affordable housing.

TRADE SECRETS

When you lighten the framing by taking out unnecessary studs and headers, you not only save forests, but also provide more room for insulation.

Look for ways to increase storage

Extra storage is critical. Take advantage of left-over nooks. I like to explore my new floor plans just after they have been framed and look for areas under stairs, next to garage walls, and between trusses to create found storage areas. Sometimes I leave these spaces insulated but unfinished so that buyers can claim them later.

Garage space is coveted

Most first-time homebuyers want an attached two-stall garage—especially in wintery climates. People thinking about moving from an apartment to a house dream about the day they can carry groceries into the kitchen without having to walk across a parking lot during a blizzard. That's why most of my homes—even some costing $75,000 or less—include at least a one-stall attached garage.

Finished attics represent a pricey but highly marketable option for providing extra living space. One option is to leave these spaces unfinished but make it easy to convert them to living space later.

Houses Are Getting Bigger

The modern affordable home has grown to an average of 1,800 sq. ft. from as little as 750 sq. ft. in the years after World War II. The Levittown homes that became a symbol of homeownership during the 1950s had only two bedrooms, one small bathroom, and an eat-in kitchen, all arranged on a concrete slab 25 ft. by 35 ft. The porches, vestibules, entry halls, and dining rooms that were standard in the 1920s were gone. These houses lacked even a basement, though some had unfinished attic space for future conversion.

But the prosperity of the next five decades presented an opportunity to recover some lost space, and new homes began increasing in size. By 1963, the average new house had nearly doubled to 1,450 sq. ft. During the 1970s, half of all new homes included two bathrooms or more.

Bedrooms were once uniform in size. Then it became common for kids to have their own bedrooms and for their parents to share a larger master bedroom. The advent of television led to the construction of family rooms for informal recreation. Increased consumerism created the need for more and more storage space to warehouse our clothes, toys, and gadgets. Eventually, room-sized walk-in closets became a necessity.

But by the early 1980s, the median price of a new house began to exceed what a median-income family could afford. The split between our dreams and our pocketbooks has continued to grow.

In this difficult climate of high expectations and low affordability, the entry-level homebuilder has to balance his or her customers' dreams with their limited purchasing power. The object of interior analysis is to provide efficient, practical living space while creating comfort and a touch of luxury. The goal is to make an affordable and livable home that your 21st-century buyer can fall in love with—not just settle for.

Space for future use

You'll find that buyers are much more willing to sacrifice certain features when they see the potential to add them later, especially buyers with more time and energy than money. Make it easy for them by designing future expansion right into your plans.

One option is to extend plumbing for a future bath under the basement slab, or rough in plumbing for a future master bathroom in a bedroom closet. You can include framed basement partitions with daylight windows to accommodate future bedrooms, or you can include a header for a future stairway to the attic in the ceiling framing so the buyer can add an attic storage area or a bedroom and bath later.

In attics, include gable-end windows (not dormers), which make the space more attractive, even if you leave it unfinished.

Other options include drafting an extra first-floor room addition right into the original plan, even if you're not going to build it. Provide enough room on site to accommodate the addition and design an easy-conversion passageway for access. Don't forget to supply future HVAC and electrical hookups nearby. Give your buyers a permit-ready set of blueprints so they can see the option on paper.

In warmer climates, design the garage for conversion to a bedroom or recreation room by providing an area for a carport in front of or beside the existing garage. Install an unfinished multipurpose loft that buyers can eventually use for an office, bedroom, or recreation area. No matter how you do it, buyers will appreciate your foresight and readily accept certain tradeoffs in exchange for expandability.

In warmer climates, converting a garage into an extra room is one way of creating low-cost living space.

Extend inner space outdoors

You can also make a house feel bigger by adding exterior "rooms." A buyer perceives a covered front porch or enclosed atrium as part of the living area, especially if you provide ample windows that help to merge indoor and outdoor areas. A 6-ft. by 8-ft. patio seems small from an outdoor perspective, but when it's used to expand the feel of an interior room, the 48-sq.-ft. addition feels huge.

Success Stories

Think Inside the Box

Beazer Homes USA of Southern California builds in one of the most restrictive regulatory environments in the United States, yet it manages to value-engineer floor plans that compete in a cutthroat real-estate market. Balancing quality and value is not just a noble goal—it's a matter of survival. A disciplined approach has helped the company become the fifth largest homebuilder in the nation. Here's how they do it:

> Balancing quality and value is not just a noble goal—it's a matter of survival.

- Beazer's marketing and design team decides on the basic shape of homes they will build in a given subdivision—either rectangular or square. Beazer's architects have learned to live within the constraints of cost-effective design, yet still produce great-looking, family-friendly homes.
- Accommodating the unique challenges of Southern California's earthquake and fire regulations can limit any builder's cost-cutting options. A value-engineering team that includes a structural engineer, a framer, a plumber, an HVAC subcontractor, and a purchasing agent find the most economical approach together.
- Purchasing agents watch the lumber markets like commodities. If they determine that lumber costs will go down, they might delay a project a few weeks to save tens of thousands of dollars.
- Inspired by Henry Ford, Beazer uses an assembly-line approach to building. For example, instead of having a foundation subcontractor trench, form, install steel, and pour one house at a time, Beazer employs a trenching company to excavate 20 homes at once, followed by forming and rebar crews, then finishers to pour the concrete.
- Sometimes you must spend money to achieve a cost-effective solution. When Beazer decided to build leakproof homes, it spent more on the shells. But warranty costs have dropped significantly.

Fact Sheet

WHO: Beazer Homes, USA

WHERE: Southern California

WHAT THEY DO: Challenged by tough building and fire codes, Beazer exploited team planning, disciplined design, and value engineering to become the nation's fifth-largest homebuilder.

Inviting outdoor spaces like a patio or deck add value and appeal to a house even when they are relatively small.

TRADE SECRETS

One of the great environmental dilemmas facing development is storm-water runoff. Roofs and pavement don't absorb water, so they create excessive flow into surrounding areas. A more natural, permeable driveway made from rock, sand, and gravel costs a lot less to build than white- or blacktop, saving you money while reducing runoff.

Lot Basics

Land planning is a key ingredient of affordability. If you buy a developed lot, setbacks and utility locations—along with their associated costs—have been determined by the time you start building. Nevertheless, proper placement of the house can offer opportunities to reduce cost.

Smaller lots present certain challenges that limit options, which is why a modular design with a small footprint helps regain some flexibility. Start with a minimum setback from the garage to the street to avoid a long driveway. If a long drive can't be avoided, consider a small concrete parking area in front of the garage with a permeable sand and rock driveway instead of an expensive ribbon of concrete or asphalt.

Use landscaping in creative ways to enhance curb appeal at minimum cost.

In general, try to avoid corner lots, which may require the construction of twice as much sidewalk. Make sure to choose a lot that does not sit too high or low, since moving dirt around can add thousands of dollars to your budget without providing any tangible benefit.

Using a flexible modular plan, or at least a right-left reversal option, try to orient central living areas toward the most favorable outdoor views. Conversely, try to preserve privacy by aiming first-floor windows toward private or fenced-in spaces and second-floor windows away from neighboring yards.

The traditional practice of setting a house in the center of the lot and leaving large setbacks on all four sides greatly reduces the usability of land on both sides of the house, especially on a smaller plot. If you have room to move your house around the lot, try to squeeze it toward the front and one side yard. This reduces driveway and landscape requirements while providing more green space in back.

CREATING OUTDOOR ROOMS

Fenced-in auto space and a trellis patio create the feeling of greater size in this 20-ft. by 20-ft. cottage designed by San Francisco architect Donald MacDonald.

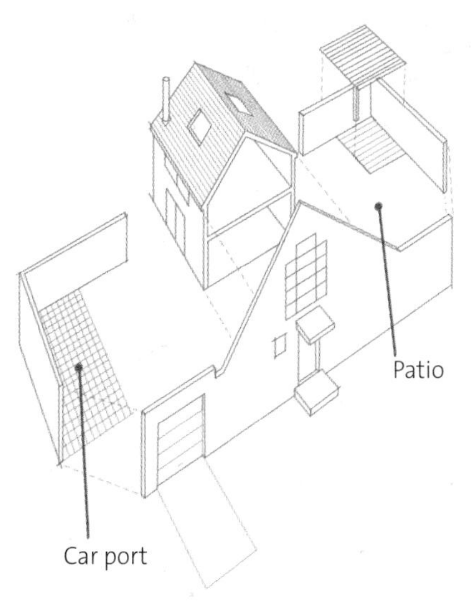

Money-Saving "Green" Checklist for Site Development

- > Minimize the disruption of existing land contours and vegetation.
- > Reduce pavement width and length.
- > Incorporate permeable paving.
- > Orient homes east and west for solar access.
- > Cluster the building site to maximize green space and reduce development costs.

Oakwood Homes in Denver sets houses 3 ft. from the side-yard property line and provides an exclusive-use easement of 7 ft. on the neighboring lot. Every house has 10 ft. of usable side yard.

To reduce site-development costs, cluster utilities into one service entry whenever possible. Keeping utilities close together also reduces construction costs.

MORE ROOM ON SMALL LOTS

The standard approach of centering a house on the lot provides two narrow side yards. By crowding the house against one property line at the minimum setback, or with no setback at all, you provide a more useful side area to enhance the value of a smaller lot. In the same way, setting the house as close as possible to the street provides a larger backyard.

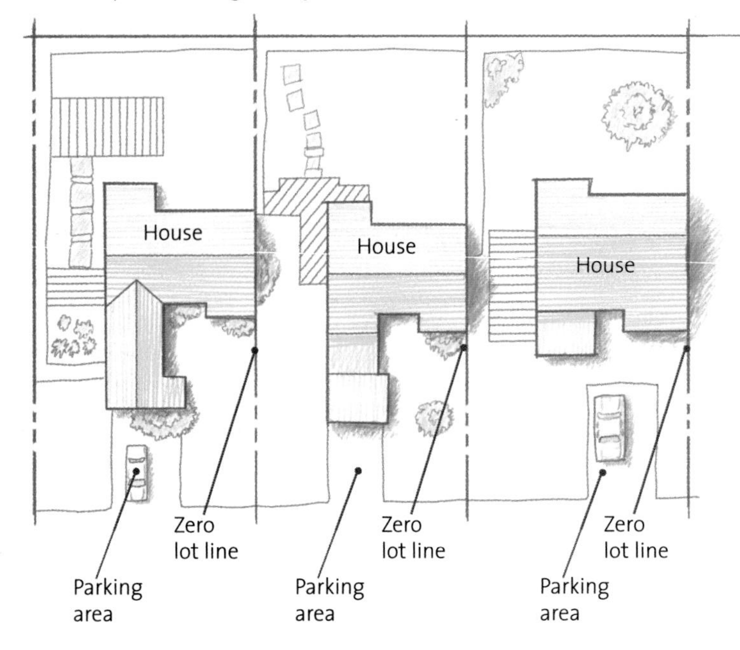

Whenever possible, coordinate utility feeds in one trench and locate meters in one multipurpose utility room. Builders who are putting up two houses on adjacent lots might consider a common driveway. Sharing infrastructure access allows you to exploit flag lots and tight site conditions that other builders often shun.

Use landscaping in creative ways to enhance curb appeal at minimum cost. A few bushes, annual color, and trees, especially, can transform a plain-Jane house into a delightful attraction. Small lots landscaped across the lot line appear large and create a parklike effect.

By exploiting natural landscape features and developing attractive neighborhood layouts, I have managed to build pockets of desirability in otherwise blighted areas. While the old real-estate adage still has validity, you can create "location, location, location" through careful site design and development in almost any neighborhood.

BUILDER'S CORNER

Working with Your Subs

Once you've succeeded in finalizing a highly marketable and cost-efficient floor plan, you're almost ready to start building. But there's one more step. The blueprint for an affordable home should work like a paint-by-number drawing, leaving nothing to chance while facilitating the work of every tradesperson. At this point, you'll have to go over your drawings repeatedly to make sure that every critical dimension has been noted clearly and that any deviation from standard construction is illustrated thoroughly.

Despite your best efforts, as soon as you start building, you'll discover that human error creeps in to waste money. This happens in almost every business, but in construction the number of people involved exacerbates the problem. It usually takes about 50 different tradespeople to build a single house, and even if your key subs remain from year to year, their helpers come and go. The cost-saving techniques you worked out with one tradesperson won't translate to the next if you have not documented them on your plans.

This is an ongoing process, even after you construct several generations of a single plan. Keep an original master plan in your office. Every time you make a modification, go back to the drawing board and record it. Every time you notice an error, fix it on paper as well as on site.

Note critical dimensions in bold lettering so subs don't have to scale off blueprints.

To avoid confusing your tradespeople with cluttered blueprints or insufficient instructions, follow commercial planning guidelines and provide each tradesperson with a separate sheet to work from. On this trade-specific floor plan, you can include everything from layouts and isometric drawings to material lists and finish schedules where practical.

Of course, well-annotated plans with complete layouts and details make for more exact material takeoffs and better bids from your subs. You probably won't manage to eliminate every error, but you can reduce slipups considerably by scrutinizing the two most common areas of construction lapse: bad measuring and lousy communication between trades.

To make sure that subs observe critical dimensions, such as the location of sewer and water stub-outs in a concrete slab, note these dimensions accurately in bold lettering so that your subs don't have to scale off the blueprints. For every critical detail, make sure to specify one common "pull point" for all related trades (that is, one point from which every trade measures, or "pulls" their tape).

For example, a discrepancy in the foundation can create major problems if the plumber locates his pipes by measuring off a rear wall, and then the framer comes along and frames the wall in which these pipes are to run by pulling his tape from the front wall. The pipes and the wall won't line up. Who's wrong? Neither one. But who pays to jackhammer the slab or move the framing? Probably you do.

One of the benefits of the roundtable approach to design (in which you gather all your key subs and vendors to review your blueprints) is the opportunity to discuss construction procedures as well as design issues. It's an area where you can find substantial cost savings and improve relations among your trade partners.

I once gathered my subs for a roundtable discussion to iron out trade differences. The question I posed was, "What can all the other trades do to facilitate your work?" This turned out to be the most profitable lunch meeting I ever attended. I learned that my subs included a fudge factor in their bids to allow for goof-ups. The electrician assumed his meter panel would not coincide with a clear stud bay, and so his bid included time for blocking and framing. The framers, in turn, anticipated awkward "fixes" after the electricians, plumbers, and sheet-metal trades cut up their walls. As my subcontractors discussed these problems, they collaborated on developing timesaving strategies that yielded many small but cumulatively substantial improvements in the construction process.

I kept track of my subs' suggestions and then hired a draftsman to draw them into my plan. Once the subcontractors felt confident their work could be streamlined and materials pared to a minimum, they eliminated the fudge factor from their bids and substantial cost reductions resulted—about 5% overall.

The Language of Affordability

Although trained designers can bring art to the science of affordable-home building, you're probably going to have to convince your architect to try it, and then educate him or her in the language of cost-conscious design. It's a language that requires comprehensive knowledge of every aspect of home building, from concrete to carpet. As you read the trade-specific chapters that follow, you will gain a broader perspective on the alternatives available to help bring a cost-effective floor plan to fruition.

> Price is only part of the equation. After you design the bottom line, review your plan and make sure it has a few lovable traits.

Even as you whittle away expenses, don't forget that the definition of *value* ultimately rests in the perception of the buyer. From the customer's perspective, builders often value-engineer the value right out of a home—it's those inefficient details like overhangs, arches, and cozy little nooks of "wasted space" that many buyers *value* most. To understand how your customers define value, you have to step aside and listen to their wishes—especially the impractical ones. Remember that price is only part of the equation. After you design the bottom line, review your plan and make sure it has a few lovable traits. Never underestimate the power of aesthetics. Don't overdo it, but always make a strategic investment in sex appeal.

TRADE SECRETS

Properly placed deciduous trees and shrubs can shield a house from the wind, provide shade, and build a microclimate that modulates winter and summer temperatures around a house up to 5° or 10°. Landscaping also helps improve sales.

A few well-placed plants and trees can enhance a home's appearance at a relatively low cost.

Foundations

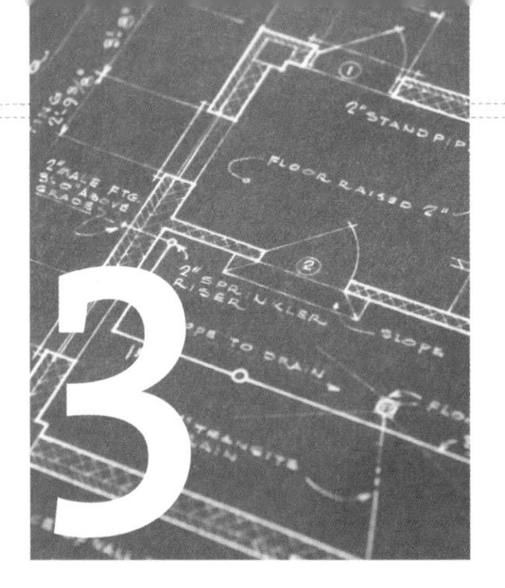

Rubble footings without concrete or reinforcing steel support the world's most ancient buildings. They still stand firmly today. In fact, when poured-in-place concrete foundations became common in residential construction in the early 1900s, builders treated the new technology with skepticism. Field-proven alternatives to poured-concrete foundations have always been available, and now these alternatives enjoy the support of modern research and engineering. Some have even found broad acceptance throughout the world in all kinds of climactic and geological conditions. But in the United States, building codes have only recently begun to adopt options like wood foundation walls, gravel footings, and precast stem walls—options that could potentially save homebuyers and builders millions of dollars every year.

Even when alternatives are unassailable, market acceptance lags behind

Even frugal builders sometimes overlook foundation costs as a gold mine of potential savings.

Money-Saving "Green" Checklist for Concrete Foundations

> Incorporate fly ash and slag in concrete.
> Use reusable aluminum form boards.
> Use recycled content rubble for backfill and drainage.

construction science, making builders think twice before trying something new. That's why I'll begin this chapter discussing techniques for minimizing costs in traditional concrete foundations while ensuring the system is adequate to the site and the structure. In the latter parts of this chapter, I'll review some of the alternate systems that prove cost-effective and have received widespread acceptance. As more builders choose these alternatives, the real-estate market becomes more willing to accept them.

> A full basement can account for nearly 8% of the total construction budget.

Concrete Is the Traditional Choice

Three factors influence the type of foundation you might choose: climate, soil conditions, and structural requirements. In warmer areas, like Florida and California, slab-on-grade foundations have become the favored method of construction. In the upper Midwest and on the East Coast, homebuyers usually prefer full basements, which provide storage and real or perceived safety in tornadoes.

TRADE SECRETS

One way to avoid soil-pressure problems on basement walls is to design shallower basements. I have used a raised-ranch basement design with only 4-ft.-high walls to satisfy the demand for basements and daylight windows while incurring lower excavation and foundation costs.

FOUNDATIONS A TO Z

Homebuyers in some parts of the country are used to seeing full basements with poured-in-place concrete walls on concrete footings. But other options cost less and will perform just as well.

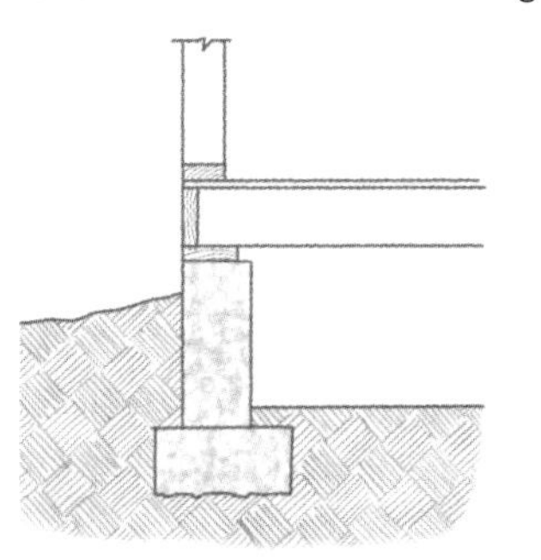

Concrete stem walls are cast directly on a gravel base, without the need for a separate footing.

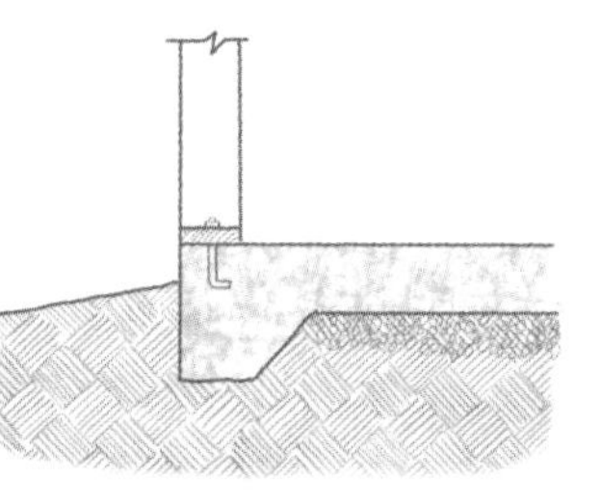

Monolithic thickened-edge slab combines footings, foundation wall, and floor in a single operation.

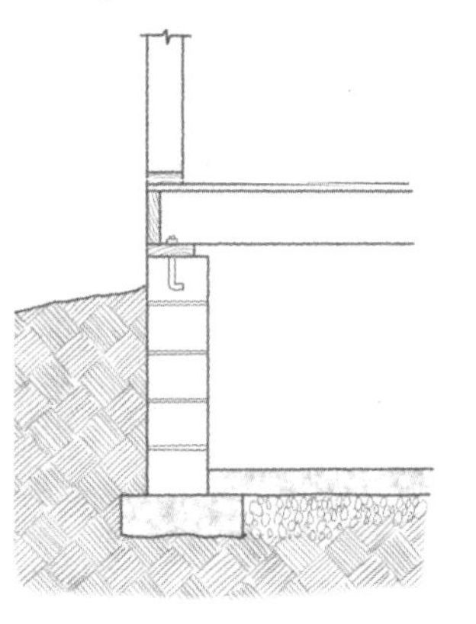

A basement or crawl space with masonry foundation wall on footing is the worst option from a cost and structural perspective.

A basement or crawl space with foundation wall on footing is a common, but costly alternative.

Depending on the water table depth, builders throughout the Midwest, the Northwest, and portions of the Southeast use both full basements and crawl space foundations.

Although slabs are not common in the upper Midwest, modern technology offers ways of coping with cold-weather footings without the expense of digging below the frost line. Frost-protected shallow footings (FPSF) and radiant-heating systems offer viable means to build slab-on-grade homes in any climate, but the tradition of building basements prevails because people are accustomed to using their homes in certain ways. If I was used to having a basement, I'd hesitate to move into a house without one. But a full basement can account for nearly 8% of the total construction budget. If I switched to a crawl space, the cost would drop to about 6%. A slab-on-grade foundation generally accounts for less than 5% of the total construction budget. But even that 5% is a sizable amount of money. It's worth the small effort to whittle that down further with value engineering. Frank Lloyd Wright's Usonian homes in Chicago, which were intended to serve an affordable-housing market in the post-Depression era, were supported on a slab without any footings (a floor slab and the gravel layer below provided bearing for exterior and interior walls). Many of these homes are still serviceable today.

The best choice for your foundation system depends on many factors, including construction time, weather conditions, the availability of skilled labor, and the cost of materials. But homebuilders usually choose a system that includes concrete, steel, and masonry because that's what buyers expect.

When using traditional materials, value-engineering options are limited to subtle improvements and innovative combinations of old technologies. With a concrete foundation, cutting costs amounts to redesigning the size and shape of the groundwork and reducing the amount of concrete and steel that's used. The opportunity to save money depends on the interrelationship of soil type, building loads, and concrete. A value-engineering team that includes a forward-thinking professional engineer, a soils expert, and your local ready-mix plant operator will help.

Match the foundation to the load it carries

The first question in value engineering a foundation is the most basic: What loads will the foundation have to support? The major building codes approach this issue with all-purpose requirements that assume below-average soil conditions and sloppy job-site quality control. A value-engineering team can design a foundation based on actual site conditions and house-specific structural requirements. You can further improve on this approach by relying on quality construction methods instead of oversizing.

A value-engineering team can design a foundation based on actual site conditions and house-specific structural requirements.

Soil conditions are especially important when building a basement because the foundation performs a double duty. It holds up the house, but it also retains the earth around it. Given the compressive strength of concrete, holding up the house is easy. A greater challenge comes from lateral soil pressure. Under identical soil conditions, every additional foot of foundation wall height increases the lateral load on that wall by nearly 50%. An 8-ft.-high basement wall holds back nearly twice the pressure of a 6-ft. wall. Consequently, shallow foundation walls require less structure than deep ones.

Even with a full 8-ft.-deep basement, an engineer can choose between a thin foundation wall reinforced with steel and high-strength

TRADE SECRETS

A local concrete plant manager probably can provide information about a wide array of concrete products that many builders don't consider because they don't know about them. To find out what's available in your area, call your ready-mix plant and set up an appointment with the manager or a knowledgeable salesperson.

BUILDER'S CORNER

Low Foundation Walls Are a Good Compromise

Although slabs are inexpensive, homebuyers in some parts of the country are used to seeing full basements beneath the house. Foundation walls less than 4 ft. high represent a good compromise in housing markets where basements are traditional. Low foundation walls are not subjected to much lateral soil pressure. In nonreactive soils, wall thickness can be reduced to 6 in., and the concrete can be placed without any steel reinforcement. These low walls can be backfilled before any floor framing goes on, reducing job-site delays and increasing convenience.

WALL THICKNESS DEPENDS ON APPLICATION

The thickness of a poured concrete foundation wall and the width of its footing varies with its application. Steel reinforcement allows thinner walls and smaller footings.

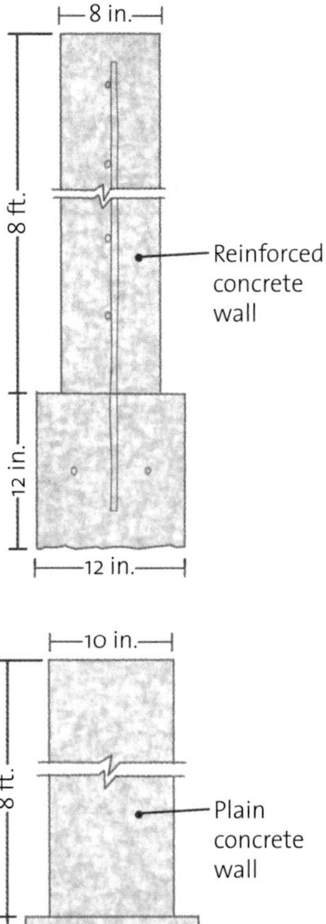

concrete, or a thicker wall of plain concrete without any steel at all. The final choice results from a painstaking comparison of costs.

Unfortunately, no single value-engineering method works for every situation. But by and large, savings of hundreds or even thousands of dollars are possible by combining optimized footings, engineered concrete mixes, reduced steel amounts, and lightweight foundation walls. A systems approach works best. Rather than treating each structural component as a separate unit, the engineer should consider the house as a single working structure. For example, if foundation walls are designed to resist the lateral pressure of surrounding soil with no consideration for the support provided by framed floors and slab, the foundation may be overbuilt.

Most builders construct all foundation walls and footings around the perimeter of a house identically. But there's no more reason to build every foundation wall exactly alike than there is to use the same size door and window headers on every opening. The section of the foundation that supports bearing walls shouldn't match the section that supports a light gable-end wall. By calculating each wall separately, the engineer can devise a more sensible foundation plan, especially for a basement with varying backfill heights, such as a walkout.

There's no more reason to build every foundation wall exactly alike than there is to use the same size door and window headers on every opening.

A Soils Test Is a Good Investment

Regardless of how much concrete you pour into the ground, the real foundation of every structure is the earth itself. For about $200, a soils technician will test the bearing capacity of the soil at your construction site. In most areas, the bearing capacity of the earth exceeds 2,500 lb. per square foot, although most codes assume 2,000 lb. per square foot or less. Armed with actual site-specific soils information and real structural loads, an engineer can usually justify a reduction in the size of footings and foundation walls.

For example, if you build a simple single-story home on silty sand (a common soil throughout the United States), the 2000 International Residential Code (IRC) requires a footing 12 in. wide. That's based on a soil-bearing assumption of 2,000 lb. per square foot and combined live and dead loads of the structure of approximately 2,000 lb. per foot of foundation wall. If the structure actually weighs half as much, and the soil has a capacity of 2,500 lb. per square foot, the footing could be 6 in. wide. In most cases, in fact, the foundation wall can sit directly on the soil without any footing at all (see Stem Wall Foundations for Residential Constructions, listed under Publications at www.huduser.org).

Even if the tests reveal trouble, such as low-bearing soil or expansive clay, it can be dealt with. Your options include replacing unstable soils, treating them chemically to change their behavior, installing drainage systems to reduce fluid pressures and swelling, and using backfill techniques that mitigate the effects on the foundation.

Coping with soil that expands

Expansive soils account for the single greatest cause of foundation failure in the United States, leading to more than $1 billion in repair costs annually. Expansive soil contains fine clay that swells when it gets wet, enough to crack concrete. It shrinks as it dries, doubling the stress on the structure through uneven settlement.

Many houses sit on expansive soil. Depending on the volume of expansive clay at your site, this problem will range from negligible to severe. In any case, it's always wise to install drainage around the foundation to keep the area dry and relieve the added pressures of groundwater.

> Expansive soils account for the single greatest cause of foundation failure in the United States.

Even when the soil is not highly expansive, fine-grained soils with low internal friction (slick soils) add substantial lateral pressure to foundation walls, especially when wet. The best draining soils, such as crushed rocks and coarse sand, have high internal friction and exert less pressure on basement walls. By backfilling with sand and gravel and installing footing drains, you can reduce lateral pressure against basement walls and thereby reduce the need to build a heavy-duty bulwark.

Slab-on-grade foundations have special problems when built on expansive soil. The conventional tactics for dealing with these problems are removing the soil or building a structure beefy enough to withstand the heaving.

In Los Angeles, where expansive soils are endemic, I've spent tens of thousands of dollars overexcavating building pads to 7-ft. depths or more, sometimes trucking in expensive fill to do so. Most foundations in the region are built with 24-in. or even deeper footings (grade beams) because of this, even when 12-in. footings could support the structure comfortably. In Texas, post-tensioned slabs are common even in affordable housing projects. Oklahoma builders employ an elaborate system of piers, grade beams, and expansive matting. In Denver, builders turn to caissons and highly reinforced foundation walls that add thousands of dollars to the cost of a house.

A more reasonable approach is to treat only the upper layers of soil that would be affected by groundwater. Instead of replacing the soil, it can be mixed with hydrated lime. The lime reduces the soil's moisture-holding capacity and improves its stability at a fraction

TRADE SECRETS

Concrete does not expand and contract significantly. The expansion joints installed between adjacent slabs or between slabs and walls simply allow differential movement. You can use a sheet of polyethylene to achieve the same end.

TRADE SECRETS

The maximum height of fill on a plain 8-in. concrete wall is 4 ft. For multilevel homes, where portions of the foundation wall do not have the lateral support of a framed floor, consider the addition of steel to prevent a hinging effect on the unsupported wall.

Soil Pressure Increases with Height

Soil pressure, like water pressure, increases with the height of the foundation wall. Each time the wall's height is doubled, the load on it increases four times. As the wall gets taller, its center of gravity also changes, improving the earth's movement arm. In other words, a tall wall provides the earth with a longer lever to pry against the foundation and tip it over.

This combination of added weight and mechanical advantage means that the substantial pressures marshaled against basement walls require a substantial structure to keep them standing. With this in mind, it's smart to build on the shallowest foundation system that is practical in the area where you live. A slab-on-grade represents the most cost-effective approach because the only loads it contends with come from the relatively light structure of the house.

COPING WITH EXPANSIVE SOIL

On sites where expansive soil can jeopardize a foundation, cutting the soil back at a shallow angle and backfilling with nonexpansive material, such as gravel, allows the adjacent expansive soils to push upward instead of horizontally against the basement wall.

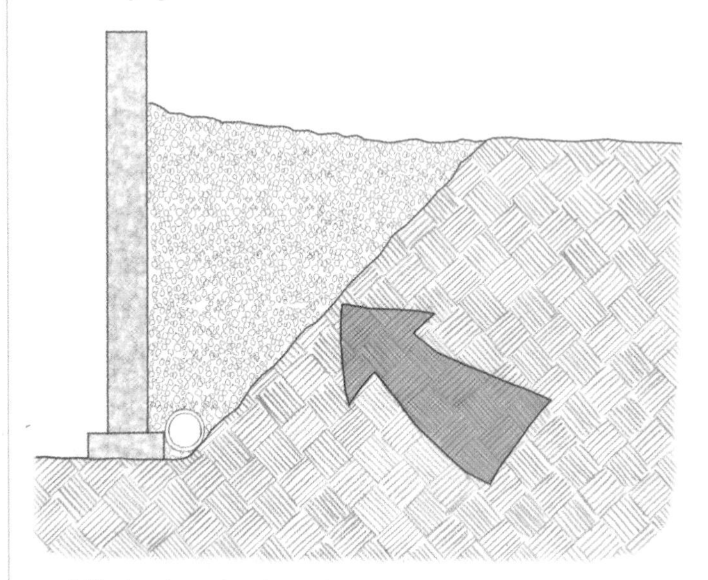

Backfilled with sand and gravel

of the cost of replacing the top layer of soil or building extra-heavy-duty footings and a hardened slab.

Another simple solution for basement foundations built on expansive soil is a sloping backfill method developed in Denver. Problem soil around the perimeter of the basement is cut back at a shallow angle, and the area is backfilled with gravel. In another approach to the same problem, two builders from Denver developed a product they call the Flor System, which provides an expansion blanket between soils and slab to absorb soil movement without heaving the concrete.

Concrete: More than the Basics

Most builders know too little about concrete to consider any alternatives to the standard mixture of portland cement, water, sand, and stone. In fact, there is a variety of concrete mixes,

TRADE SECRETS

Fly ash and slag are waste products from coal-fired furnaces at power plants and blast furnaces at steel mills. You can substitute 5% to 70% of the portland cement content of concrete with these recycled products to reduce costs while improving the strength and workability of the mix.

Ready-mix concrete plants offer more options than many builders are aware of, including additives and concrete blends that cut foundation costs without sacrificing necessary strength or workability.

Aggregates come in fine to coarse grading, such as the sand, gravel, and crushed limestone shown here. Concrete with more fine aggregate requires more cement. A higher percentage of coarse aggregate reduces the amount of cement needed—and saves money.

characterized by strength, workability, and resistance to abrasion and weather. Less expensive products like fly ash can replace a certain amount of cement, improving your bottom line and enhancing the strength and workability of the mix.

It would take an entire book to describe the options available from your local batch plant. When I checked with my ready-mix representative, I found he had formulas for 357 different concrete mixes. Then he apologized for the "limited" selection. "We're a small company," he explained. A variety of admixtures and fibrous aggregates can change the characteristics of concrete and make it much more versatile.

Larger aggregate means lower costs

You can specify concrete just by telling your ready-mix dispatcher what you're building. Most suppliers have standard mixes for basements, footings, and slabs. But to save money, you can vary the standard mix by reducing the amount of portland cement, which is by far the most expensive ingredient in concrete. Most foundations don't require concrete with a rating of more than 1,500 psi, considering that concrete blocks, often used to build basement walls, are generally made with 1,500 psi concrete, if not less. On the other hand, if your mix isn't workable and you add water, you can significantly compromise the strength and performance of the concrete. Always tell the ready-mix operator what you'll be doing with the concrete (making footings, slab, basement walls, etc.) and ask for a recommendation.

Through a chemical process called hydration, portland cement reacts with water to create an adhesive paste that coats each grain of sand and rock. Then it hardens, binding all the ingredients into a rock-solid fusion known as concrete. The smaller and more numerous the aggregates, the more cement is required to coat every single grain. In other words, the more

TRADE SECRETS

Much like a concrete footing, a gravel footing can distribute foundation loads to a sufficient soil area. It also provides drainage. Of the various gravels you can use to construct a footing, pea gravel works best because it does not require tamping and can easily be screeded to a level surface, much like concrete.

Supplementary cementitious materials, such as the fly ash on the left, strengthen concrete and add desirable working characteristics while reducing the amount of expensive portland cement (right) in the mix.

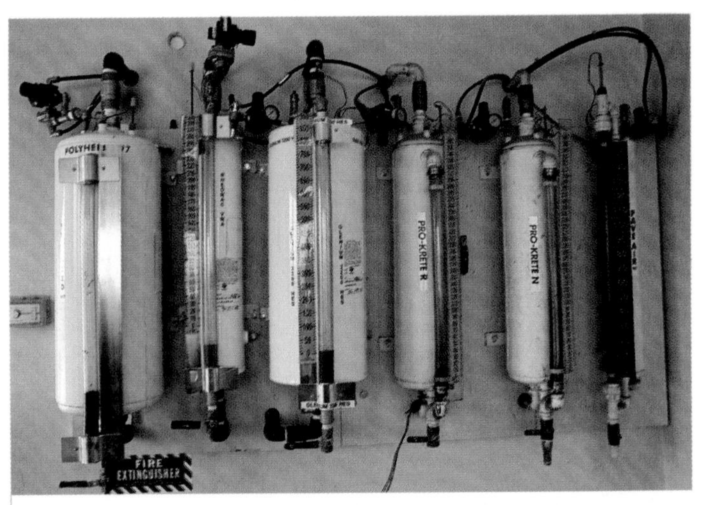

Admixtures are chemicals added to concrete to enhance its characteristics. The pumps shown add an air-entraining agent, a water-reducing agent, and accelerators to concrete as it is loaded onto the ready-mix truck.

sand and pea-sized rock you use, the more cement your mix requires. This is why you can order a ¾-in. gravel mix with four or five sacks (376 lb. to 470 lb.) of cement per cubic yard, while a finer mix using ¼-in. gravel requires at least six sacks (564 lb.). One way to cut the cost of your concrete and improve its strength is to use the largest aggregate that is practical for a given application. But, if you are pouring in tight forms or hard-to-fill places with limited consolidation (that is, no vibration), then specify smaller aggregate and a more workable mix.

One way to cut the cost of your concrete and improve its strength is to use the largest aggregate practical for a given application.

It's worth asking your foundation subcontractor what kind of aggregate he's using and why. Many use a finer mix to improve workability and avoid any exposed gravel or honeycombing on the exterior surface of basement walls. My own foundation subcontractor used to pour a mix with only 10% of ¾-in. rock. It required 564 lb. of cement at a total cost of $12.60 per yard of concrete. After some discussion with the ready-mix plant manager, we increased the rock quotient to 20%, which reduced the cement required by a half sack and shaved about $1.60 off each yard poured. This change didn't reduce the quality of our concrete, and workability never became a problem because we added fly ash to the mix. On the rare occasion that some honeycombing occurs in a visible spot, we patch it in minutes with a skim coat of mortar. Proper pouring techniques usually prevent this problem even when using coarse aggregates and low-slump concrete.

Cement substitutes help reduce costs

Ready-mix plant operators can add other materials to reduce the cost of concrete. Commonly described as "mineral admixtures," some of these materials, called *pozzolans,* react with portland cement to form cementitious compounds.

The most common mineral admixtures for residential construction include ground granulated blast-furnace slag and fly ash. Both of these can have cementitious properties that enhance the strength and workability of concrete while reducing the amount of portland cement required in the mix. Fly ash can replace between 5% and 65% of the portland cement in concrete, slag from 20% to 70%.

If you consider that portland cement costs about $80 per ton and fly ash or slag about $20 per ton, you can see why the admixtures save money. They also make the concrete more durable. The only drawback is longer setting times, which can present a problem when pouring in cold weather. You can reduce the percentage of supplementary materials in winter or add chemical admixtures that accelerate the setting time.

Chemical admixtures also offer savings

Chemical admixtures are natural and manufactured additives that enhance the workability, durability, and strength of concrete. The most common admixtures for residential construction include air-entraining agents, water reducers, retarders, and accelerators. Air-entraining admixtures help concrete resist damage from freezing by adding millions of air bubbles that act as tiny expansion chambers, allowing concrete to handle freeze and thaw cycles without fracturing. Water reducers temporarily increase the slump of fresh concrete to make it easier to pump and finish (lower water-to-cement ratios make concrete stronger). Retarders slow the set time to facilitate warm-weather pours, while they enhance cured concrete's strength. Accelerators speed the set of concrete so you can pour, finish, and insulate it before freezing temperatures set in.

A slab is the most economical option available for a residential foundation. Insulating techniques allow them to be used successfully in all parts of the country, including regions with long winters and very cold temperatures.

From a cost perspective, water reducers and super-plasticizers (high-range water reducers) allow the mix to have less water, larger stone aggregate, and less portland cement without compromising workability. This also yields a very high-strength concrete that can exceed code standards. Normal water-reducing agents improve slump by about 10%, while super-plasticizers can improve it by 25% to produce a flowing concrete suitable for walls even when using a 20% stone mix. The savings comes from using less cement and larger aggregate.

TRADE SECRETS

Finishing fiber-mesh-reinforced concrete takes some getting used to. It appears stiffer than regular concrete, but this does not affect its workability. It's critical not to add water to the mix to improve workability because added water can seriously compromise the strength and durability of the concrete. In addition, wait for bleed water to evaporate. Finishing too soon will pull fibers onto the surface of the slab and make it more difficult to trowel.

Fibers can replace wire mesh and rebar

One of the most useful concrete enhancements available today comes from a variety of synthetic and steel-fiber additives. Blended with standard concrete and requiring no extra labor, man-made and metal filaments can provide benefits comparable to, and sometimes exceeding, those of standard wire mesh and reinforcement bar. In most residential applications, they also improve the quality of concrete to provide a more reliable finished product at a substantially lower cost.

In most residential applications, the switch to fiber reinforcement saves time and money.

Synthetic fibers specifically engineered for concrete can help to reduce cracks, make concrete less permeable to water, and provide a hard surface that resists abrasion and spalling. Certain types of synthetic fiber have tested as equal to or better than welded wire mesh in nonstructural slabs. In certain instances, steel fiber and blends of steel and synthetic filaments can replace reinforcement even in structural concrete. However, you should always make a cost comparison because steel fiber can cost more than steel in heavily reinforced concrete.

Look for engineered synthetic fiber

Although you'll find many different brands of synthetic fiber, always specify a brand-name fiber specifically designed for concrete reinforcement, such as Fibermix® from Synthetic Industries®. Most fibers are filaments of polypropylene, a common plastic, so some manufacturers use recycled products such as polypropylene rope to make their concrete additives. While recycling is generally a good thing, these products don't perform as well as those specifically designed for concrete because they lack the physical properties and special lubricants that force fibers to spread throughout the mix.

The benefit of fiber comes from its even distribution. Uniform fiber distribution helps suspend concrete aggregates to prevent problems like honeycombing, microshrinkage cracks, and the development of large capillaries. There are two kinds of fiber available at most ready-mix plants: monofilament, which looks like pieces of fishing line (Fibermix), and fibrillated fiber, like Fibermesh®, which comes bound in little packets that pull apart like a row of paper dolls.

Fibrillated fibers break into millions of multidirectional strands of reinforcement that can actually bridge cracks and disperse the stresses on concrete better than steel mesh. Using it can save about 30% over concrete with wire mesh. This savings assumes you use the old and ineffective "hooking" technique to install your steel mesh. If the wire is set up correctly, so it is suspended on masonry chairs and floats in the upper third of the slab, the cost benefits of using fiber additives are even greater.

Steel fibers for structural slabs For footings, walls, and structural slabs, steel filaments provide a tested alternative to steel wire and reinforcement bar. This is especially useful for

Stronger Homes for Less

After Hurricane Andrew ripped through southern Florida in 1992, building codes became understandably stricter, prompting a dramatic increase in building costs and lengthening the time it took to build a house. Most homebuilders accepted the changes and raised their prices, but Mercedes Homes of Melbourne, Fla., searched for alternatives.

The most expensive parts of building hurricane-resistant homes are exterior block walls. Block takes longer to erect than conventional wood framing, costs more, and creates insulation and finish challenges. Mercedes Homes experimented with insulated concrete forms, tilt-up wall products, and precast wall products. All had their advantages, but they generally cost more than traditional methods of construction.

Kirk Malone, Mercedes Homes's corporate vice president of construction, saw a potential breakthrough in the aluminum forms used in some parts of the country for poured-wall basements. Mercedes Homes fielded a crew of basement contractors to build an experimental concrete home in Melbourne. The experiment led to a permanent division at Mercedes Homes that now pours about three solid-wall concrete homes a day. These walls are 30% more energy efficient than block walls and much stronger.

These walls are 30% more energy efficient than block walls and much stronger.

The exterior is finished with color-coat stucco, tooled to create affordable but expensive-looking details around windows and entryways. Inside, Mercedes Homes uses 25-ga. framing, which two people can erect in two days. Mercedes Homes collaborated with architect Steve Winter to develop an innovative roof truss design that allows the builder to contain the air-handling system completely within the building envelope. The finished structure is resistant to fire, wind, and insects, and Mercedes Homes has managed to maintain a competitive price advantage.

Fact Sheet

WHO: Mercedes Homes

WHERE: Melbourne, Fla.

WHAT THEY DID: Adapting techniques used to form basement walls, the company pioneered cast-concrete houses that stand up to hurricanes and cost less to build than concrete-block houses.

poured-in-place walls, where installing horizontal and vertical steel bars can take several days. Synthetic Industries has obtained code approval for a blend of steel and synthetic fibers called Novomesh ICF® as a substitute for traditional reinforcement of insulated concrete forms above grade. For now, below-grade application requires the approval of a structural engineer.

Some contractors have found that a blend of steel and synthetic fiber mesh provides a good alternative to traditional reinforcement for slabs poured over mildly expansive soils. This works especially well in stamped concrete applications, where wire can interfere with the finishing process.

Steel and synthetic fiber blends also can be used in combination with traditional reinforcement to reduce the steel required in certain assemblies, such as post and column connections or grade beams and piers. Fibers work well in these applications because excessive, tightly bound steel can result in more difficult placement of concrete and inadequate coverage over some reinforcement areas.

Synthetic Industries Novocon® steel fibers and high-performance S-152 HPP polymer concrete reinforcement are used to strengthen concrete walls and slabs.

In general, the savings in labor and material for steel-and-fiber-reinforced concrete are substantial in walls and slabs up to 6 in. thick. Beyond this width, steel fibers don't provide a competitive advantage because adding enough of them to suffuse thicker concrete costs more than tying steel. But in most residential applications, the switch to fiber reinforcement saves time and money. The builders I've spoken to who have made the change don't regret it.

Do You Really Need a Footing?

The footings and foundation for a typical one-story home usually don't have to support more than 1,500 lb. per foot, and often less than 500 lb. per foot. Loads on two-story homes generally range from 1,000 lb. to 2,000 lb. per foot of foundation. When you consider that a typical 12-in. square footing can handle loads of more than 40,000 lb. per square foot, it becomes obvious that most residential footings are overbuilt. Most soil can bear from 2,500 lb. to 5,000 lb. per square foot so there's little point

Millions of dollars are buried every year in oversized and sometimes unnecessary footings. In some situations, footings can be eliminated completely without jeopardizing the structural integrity of the house.

TRADE SECRETS

Foundation-grade treated lumber is not the same as deck lumber. The preservatives in ALSC-accredited foundation-grade lumber are not water-soluble and, after being kiln-dried twice, pose no risk to the environment.

Footing Width for Typical Loads

	Soil-bearing pressure (lb. per sq. ft.)			
Design load per lin. ft. of footing	**1,500**	**2,000**	**2,500**	**3,000**
1,000	8 in.	6 in.	5 in.	4 in.
1,500	12 in.	9 in.	8 in.	6 in.
2,000	16 in.	12 in.	10 in.	8 in.
2,500	20 in.	15 in.	12 in.	10 in.

in designing a footing that can hold 10 times more weight than the ground underneath it.

The primary objective of a footing is to provide a level worksurface on which to build. Its secondary objective is to place building loads onto solid bearing, below surface organic soils that tend to be porous and weak. On rocky ground or firm soil, footings may not be necessary at all. In fact, until the early 20th century, most homes rested right on the surface of the ground—and many of these homes still stand today. More recently, research sponsored by the U.S. Department of Housing and Urban Development (HUD) and demonstrated on a variety of housing styles and sizes has shown that concrete foundation walls can be placed directly on soil or a leveled gravel footing.

It's hard to believe, but traditional concrete footings usually aren't necessary—they're just built from habit. For example, if a concrete foundation wall is built on stable ground and deep enough to avoid frost, an extra 12 in. of footing serves no structural purpose. Generally, the foundation wall acts as a giant grade beam providing all of the soil-bridging capacity that's needed. This is why most of the modern engineered foundation systems, like precast concrete, wood foundations, and cast-in-place stem walls, rest on a bed of gravel. Gravel provides a level surface and good drainage.

Even if the soil requires the type of bridging that footings provide, or if you are building in a seismic area, a value-engineering approach can yield substantial savings. For example, residential structures generally do not require the two no. 4 reinforcement bars that most builders install along the length of their footings unless the footings are extremely shallow. Even then, if the subgrade is prepared carefully, you can generally eliminate the steel. In fact, most codes do not require steel if the outer surface of the foundation walls are equal to or less than the depth of the footing. In other words, in good soil, if you're placing an 8-in.-wide foundation on 8-in.-deep footings, you can safely skip rebar without consulting an engineer.

In any event, placing horizontal steel in a relatively thin footing is the least advantageous use of steel. The steel creates much greater leverage and strength against differential settlement when it is placed horizontally at the top and bottom of the foundation wall.

Again, the footing (if there is any) only needs to provide a level worksurface on which to build the wall and, in weaker soil conditions, to distribute the bearing load on a wider surface—in much the same way as gravel. Only in the most heavily loaded and weak soil conditions is footing reinforcement required.

Special situations

Where soil conditions or heavy loads require extra support, you can reduce the thickness of the footings by adding reinforcement. This usually consists of horizontal bars that don't require a lot of labor to install and often represent a less costly alternative to digging deeper

TRADE SECRETS

Soils containing silt are the most frost-susceptible materials to build upon. Relatively clean sand and gravel generally don't heave, because they don't allow water to be readily transported from below to the freezing ground above by capillary action. Sites that are dry or well drained generally do not have frost-heave problems either, even if the soils are silty, because there is no available moisture to wick up to the freezing ground, where it can accumulate as frozen layers or ice lenses that jack the soil (and anything on it) upward.

POUR INTERIOR FOOTINGS WITH SLAB

Instead of digging interior footings and pouring them along with perimeter footings, wait until after you backfill with granular subbase and then hand-dig interior footings and pour them monolithically with the floor slab.

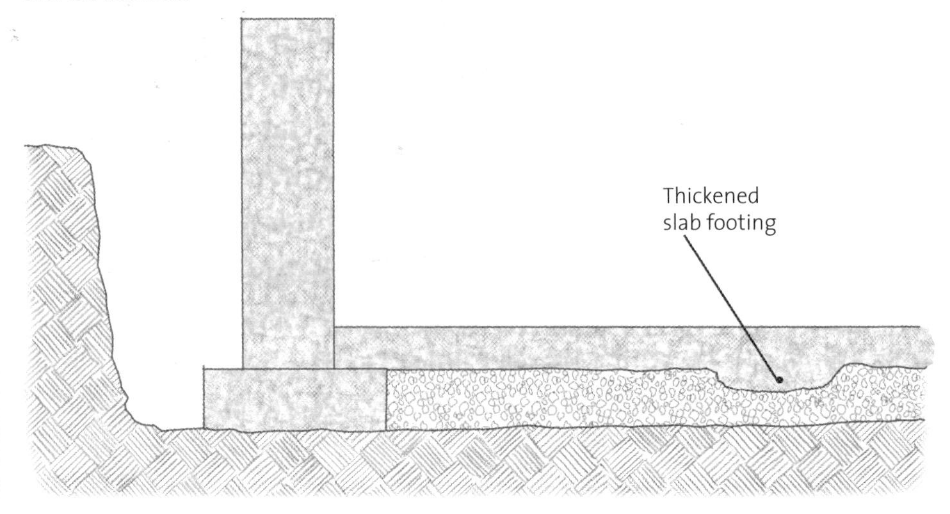

TRADE SECRETS

Don't size your footing widths by available backhoe buckets. When preparing narrow footings, have your backhoe operator overexcavate footings, then backfill the trench with sand and gravel. Take a square-nosed shovel and dig footings to the exact dimensions designed by your engineer. Digging through sand and gravel is easy, and you'll save money on concrete while benefiting from the added drainage provided by granular fill. Another solution is to use a trencher instead of a backhoe whenever possible. You can rent this tool for a day and dig clean, narrow footings that don't consume extra concrete.

and using more concrete. But bear in mind that a typical ½-in. reinforcement bar has a tensile strength of more than 60,000 psi. Use rebar only when it offsets the cost of additional concrete. Like concrete, steel can add structure far beyond the needs of a lightweight wood-framed home.

Use flowable fill for deep trenches If you have to dig through several feet of soft soil to reach good bearing, don't use expensive concrete as trench fill. Order a load of gravel instead, and fill the trench the old-fashioned way—with rock. Another option comes straight from the ready-mix plant: flowable fill. Sometimes erroneously called slurry mix, flowable fill is a low-strength concrete viscous enough to be self-leveling. It sets to a consistency of hard sand and has the bearing capacity of compacted soil, or about 150 psi. The ready-mix plant delivers flowable fill in a truck, just like concrete, but it costs about a third as much.

> Once an engineer calculates actual structural loads, you might be able to replace 12-in. interior footings with 6-in. footings only 4 in. deep.

Make interior footings smaller Builders usually construct interior footings (used to support columns in a basement, for instance) using the same dimensions as exterior footings. But interior footings typically support half the load and are contained within the dry and heated envelope of the structure. Once an engineer calculates actual structural loads, you might be able to replace 12-in. interior footings with 6-in. footings only 4 in. deep.

Even without re-engineering your foundation plan, you can save money on interior footings by measuring their depth correctly—from the top of the slab. When your plan specifies a 12-in.-deep interior footing, you should only dig 8½ in. into the soil because the slab will provide the other 3½ in. of concrete. If your basement rests on a firm, natural base, it's less expensive to scratch cavities for interior footings into the gravel sub-base and pour them at the same time as the slab. This helps your plumber, too, who does not have to deal with interior footings blocking his path when laying out his soil lines.

A separate, properly sized footing is required only when an interior wall exceeds 1,500 lb. per foot. Many interior partition walls weigh about 50 lb. or less per foot. A standard 3½-in. slab can support loads up to 500 lb., making a separate footing for these lightweight walls unnecessary. When building on mildly expansive or soft soils, you can add one reinforcement bar running longitudinally in the slab directly underneath the wall. For interior partitions weighing 500 lb. to 1,000 lb. per foot, thicken the slab by just 2 in. and tie 24-in.-long steel bars perpendicular to it on 24-in. centers. If your wall weighs 1,500 lb., thicken the slab by 4 in. Anything more than 1,500 lb. requires a separate footing.

Foundation walls may not need steel

Just like footings, cast-in-place concrete foundation walls don't always require steel reinforcement. In lieu of steel, you can generally thicken your walls or build them with higher-strength concrete. When you take into account the lateral support provided by floor framing, an 8-ft.-tall, 8-in.-thick plain concrete wall, with no steel at all, can support up to 7 ft. of fill. If you're concerned about temperature and shrinkage cracks, just one no. 4 reinforcement bar run horizontally along the top and bottom of the wall can minimize any separation and make the wall act as a deep beam to prevent differential settlement.

Tying a grid of horizontal and vertical steel into a foundation wall is both time-consuming and expensive. This is why I recommend looking into a plain wall construction with reinforcement, as described above, to control cracking and to gain the most strength against differential settlement at the least expense. The IRC provides a table listing the maximum wall and backfill heights for various wall thicknesses in different

High-precision, lightweight aluminum forms for concrete foundation walls make building basement forms quick and precise work compared with old-fashioned plywood forms.

Foundation Wall Thickness

Foundation walls with backfill on only one side are considered unbalanced. The higher an unbalanced wall, the thicker the wall must be. These numbers apply to a foundation on soil that will support between 2,000 and 3,000 lb. per square foot.

Maximum wall height	Maximum unbalanced backfill height	Plain concrete nominal wall thickness
5 ft.	4 ft. to 5 ft.	6 in.
6 ft.	4 ft. to 6 ft.	6 in.
8 ft.	4 ft.	6 in.
	6 ft.	8 in.
	8 ft.	10 in.

TRADE SECRETS

You should always make a special effort to ensure a square and level foundation. But to save time and trouble when your best intentions don't work out, undersize the foundation by 1 in. overall. Then frame your walls to the proper measurements. The miniscule difference won't be noticed, but it can save you hours of head-scratching and dimension-adjusting as you try to compensate for a forming error.

Soils Support the Weight

Generally, you'll encounter two types of soils on your construction sites: coarse-grained sands and gravel, and fine-grained silts and clay. You can make a quick appraisal of the soil you're building on by picking up a handful and looking at it carefully. If you can readily see the particles, chances are you're looking at sand or gravel. This is the best building soil because it drains well and provides excellent support for footings.

In contrast, fine-grained silts and clay-type soils tend to clump together and don't separate into individual grains. Wet a handful of soil and then mold it into the shape of a cube. Using finger pressure, break the cube in two. If it breaks with a pop, or if it's difficult to fracture, then most likely you're dealing with clay-type soil. If it breaks easily or falls apart under light pressure, then it's made of organic matter.

To test the clay content further, try rolling the soil into a thin thread about 1/8 in. in diameter. If it cracks, or if you can't roll it at all, then your soil has low clay content. If it can be rolled into a smooth thread, try rolling it into a ball. Easily malleable, moldable soil indicates high clay content and the potential for trouble. Call a soils engineer and have the site tested further.

Load-Bearing Values for Soil

Type of soil	Load-bearing pressure (in lb. per sq. ft.)
Bedrock	12,000
Sedimentary and foliated rock	4,000
Sandy gravel and gravel	3,000
Silty sand, clayey sand, and silty and clayey gravel	2,000
Clay, sandy clay, silty clay, clayey silt, silt, and sandy silt	1,500 (call the soils engineer)

TRADE SECRETS

When building a plain concrete basement wall, it's wise to add a ½-in. steel reinforcement bar near the top of the wall to create a bond beam, as well as over any openings, such as windows.

soil classes. Study this table and see if you can build your standard foundation walls with minimum steel. You'll find the savings can add up to about $600 in a typical basement foundation.

Whether you use steel or not, make sure foundation walls are no bigger than they have to be. There's no need for excess structure when standard wall construction already represents one of the most conservative and overbuilt assemblies in a home.

Whenever excess structure is reduced, however, quality control has to increase. In this case, avoid backfilling against a plain (unreinforced) concrete wall until you install both the basement slab and the floor framing. Although this rule actually applies to all basements, even when reinforced, it becomes especially important when you use plain concrete. Most basement cracks occur immediately after construction because builders backfill too soon, and the uncured concrete cannot handle the load. It's rare that a foundation fails because of too little steel or inadequate concrete. Contrary to the opinion of many contractors, beefing up foundation walls won't prevent cracking nearly as efficiently as careful construction.

Consider Alternatives to Conventional Foundations

Alternatives to conventional foundation walls include stem walls, treated-wood foundations, precast panels, and insulated concrete forms. Of these, treated-wood and precast panels represent the best options for cost control.

In some areas, these products aren't locally available, and the added cost of shipping bulky and heavy construction supplies makes it hard for the alternatives to compete with regional construction methods. In contrast, stem-wall construction is one alternative system for basement and crawl-space walls that can be built with standard, locally available materials and common knowledge of concrete work.

Monolithic stem walls have no footings

Stem-wall foundations are built with walls engineered to distribute building loads directly onto soil without a separate footing. They are available as precast products or poured on site. The on-site construction of a monolithic stem-wall foundation is similar to that of a conventional concrete wall, except that by setting the wall directly on natural soil or a gravel bed, you save approximately $9.50 a foot for trenching, forming, and concrete. On average, this translates into a savings of $500 to $1,000 per house.

Although the stem-wall technique is already prevalent in slab-on-grade construction (where a thickened slab edge rests directly on the earth and doubles as wall and footing), the major codes do not offer prescriptive guidelines for stem-wall design when applied to basement and crawl-space foundations. This might make it necessary to obtain an engineer's stamp to support the stem-wall concept with your local building department. An engineer might charge $250 to $500 for this, making this approach most useful for builders who use the same plan time and again.

Stem walls are essentially concrete grade beams 8 in. to 10 in. thick. Their thickness varies with loading conditions, wall height, house width, and the bearing capacity of the soil. Eight-inch walls represent an economical alternative for most residential structures on good soils, and the 10-in. walls provide an option for less-than-ideal sites or heavy structural loads.

You can excavate a basement and build these walls using standard poured-wall forms, or you can trench stem walls into the ground for a crawl-space foundation. The system requires either a concrete slab to secure the toe of the wall or 12 in. of soil on either side. Two ½-in.-thick reinforcement bars run horizontally, one along the bottom of the wall and one along the top, providing bridging capacity over any soft spots in the soil.

HUD and the National Association of Home Builders (NAHB) Research Center have published a design guide called Stemwall Foundations for Residential Construction (NAHB Research Center, Upper Marlboro, Md., 1993, Instrument No. DU100K000005897). This booklet contains easy-to-read tables that you or your engineer can use to plan your foundation.

Treated-wood foundations gain converts

Treated wood is another proven cost-effective material for basement and crawl-space foundations in areas of the country without a high water table or extreme termite problems. In parts of the upper Midwest, in the Southwest, and throughout Canada, these foundations have gained broad market acceptance because they provide warm, dry basements that can be finished more economically than concrete or masonry block.

Structural insulated panels made with foundation-grade plywood for warm, superinsulated basement walls can be installed quickly.

A permanent wood foundation provides a cost-effective, all-weather alternative to concrete.

Wood Foundations Have a Reliable Record

Foundation-grade treated wood is treated and kiln dried to a retention level of 0.60 lb. of preservative per cubic foot of wood. All wood used in permanent wood foundations should be identified by a quality mark of an inspection agency accredited by the American Lumber Standard Committee.

Very few problems have been reported on the more than 200,000 wood foundations installed in the United States over the last 50 years. Problems generally occur when builders use the wrong materials or don't follow industry guidelines. This contrasts sharply with concrete. Surveys by the U.S. Department of Housing and Urban Development (HUD) and the National Association of Home Builders (NAHB) reveal that nearly half of all homeowner complaints relate to concrete.

Site preparation and excavation procedures are the same for a treated-wood foundation as for a conventional system. As with all foundations, footing sizes depend on the height of the foundation wall, lateral soil pressure, and building loads. You can use concrete footings, but either gravel or crushed stone works better because it provides drainage. Good drainage is critical for every type of basement, but especially for one made of treated wood.

Because a wood foundation depends on floor framing to resist the lateral pressure of backfill, any eccentricities in the structure, such as stairways adjacent to the perimeter wall or openings for doors and windows, require special consideration. But several books are available that provide such detailed design and construction information, and you'll find treated wood much more forgiving and less expensive to repair than concrete.

A related method for permanent wood foundations involves structural insulated panels made with foundation-grade plywood and studs. This system provides excellent insulation and relative ease of construction. Holes in the plate allow for the installation of electrical wiring. Although more expensive than traditional wood foundation systems, insulated panels work well in cold climates.

Citation Homes, in Spirit Lake, Iowa, built one of the first wood foundations in the United States. After nearly 50 years in the ground, paint has started peeling below the downspout, but the foundation walls show no signs of deterioration.

After installing and leveling a footer plate, the builder will snap lines for walls and begin framing.

Foundation walls running parallel to floor framing require two bays of blocking for lateral support.

Although the Empire State Building in New York rests on a wood foundation, and wood foundations have stood the test of time in many parts of the world, some buyers in the United States still don't like the idea of a perishable product holding up their houses. But treated-wood stakes set in swamps nearly 70 years ago have not deteriorated yet, and some manufacturers of wood-foundation systems guarantee their products for 50 years—try getting that kind of a guarantee from your concrete supplier. I use treated-wood foundations frequently, especially on rental homes where marketing concerns are not important. They are also great alternatives for light-commercial buildings.

By using a wood foundation, at about $35 a foot, I can save an average of $1,500 per house over the cost of a conventional basement

By using a wood foundation, at about $35 a foot, I can save an average of $1,500 per house over the cost of a conventional basement, which runs nearly $55 a foot. Wood foundations don't freeze, so it's possible to build them throughout the winter. I can also build a wood foundation with my own carpenters, which eliminates the hassles of coordinating trades and fitting into a subcontractor's busy schedule. They also permit easier finishing and wiring.

But treated wood provides significant cost savings over concrete only when used to build crawl-space foundations and simple basements with backfill of 84 in. or less. In complex configurations with uneven backfill, such as walk-out basements, or where backfill exceeds 84 in. in depth, the structural framing required can offset the cost benefit of using wood.

Although some builders frame their own treated-wood foundations on site, I prefer to have them built by an experienced wall-panel plant. The concept of a wood foundation is simple—almost like framing a conventional wall below grade—but not exactly. Because below-grade wood-framed walls have to resist lateral soil pressures as well as the vertical loads of gravity, it's critical to follow the proper framing, fastening, and blocking techniques. Unless you're very experienced with wood-foundation techniques, it's best to have qualified pros design and fabricate your walls.

Like any construction system, the specifics of designing and building a wood foundation depend on the soil conditions at your building site and the structural requirements of your home.

Precast foundations are made off site

Precast concrete wall panels that are assembled into a ready-made foundation system have become popular in certain markets, especially along the East Coast. If your building site is

Two men and a small crane install a Thermal-Krete® foundation in one day.

close enough to a manufacturing plant, precast concrete can provide an attractive and reasonably priced alternative to a conventional foundation. But shipping costs make this choice less attractive if the job site is more than 100 miles from the plant. And unlike wood panels, precast concrete panels require a crane on site. Despite these drawbacks, many builders swear by concrete panels as the most expedient and economical foundation alternative they have tried. Prices vary around the country, ranging from $50 to $70 per foot, including installation.

The major benefits of using precast panels include speed and convenience. A typical installation requires one day to dig the hole and prepare the gravel bed and one day to set the walls with a two-person crew and a crane. This compares very favorably to a five-day, five-person poured wall or concrete-block system. Weather doesn't affect the panels, so you can set them at any time of the year. I spoke with one installer from Needham, Mass., who had dug and installed a precast foundation system during a snowstorm. The next day, the builder placed a modular home on this foundation. The storm had paralyzed construction on every other building site in the city for more than a week.

Superior Walls of America® is the largest manufacturer of code-approved walls with fabrication plants in 23 states. Their panels are cast like bond beams with an integrated footer board, which means these walls don't require a separate footing. Steel-reinforced concrete studs provide room for insulation and ready application of drywall without furring. Although these walls weigh less than half as much as a concrete or masonry foundation, they are made of high-strength, fiber-reinforced concrete that provides sufficient bearing for almost any residential structure. A factory-applied sheet of Dow insulation provides an R-5 value for their standard wall, and up to an R-13 value for their upgraded panels.

Ordering precast concrete walls is very similar to ordering trusses. A builder sends plans to the factory where walls are designed and cast, then shipped to the site. Unlike trusses, which are erected by the builder, precast concrete panels are set by a certified installer. All the builder does is dig the hole, level a 6-in. bed of gravel at the base, and provide the installer with two reference points for laying out the foundation. Just like treated-wood foundations, precast walls rely on a basement slab and a framed floor system for lateral support.

PRECAST CONCRETE FOUNDATIONS SAVE TIME

Precast panels are made like bond beams, with an integrated footer board that distributes the house and foundation load at a 45° angle through the gravel bed. It's the equivalent of a 16-in.-wide footing for a 10-in.-thick wall.

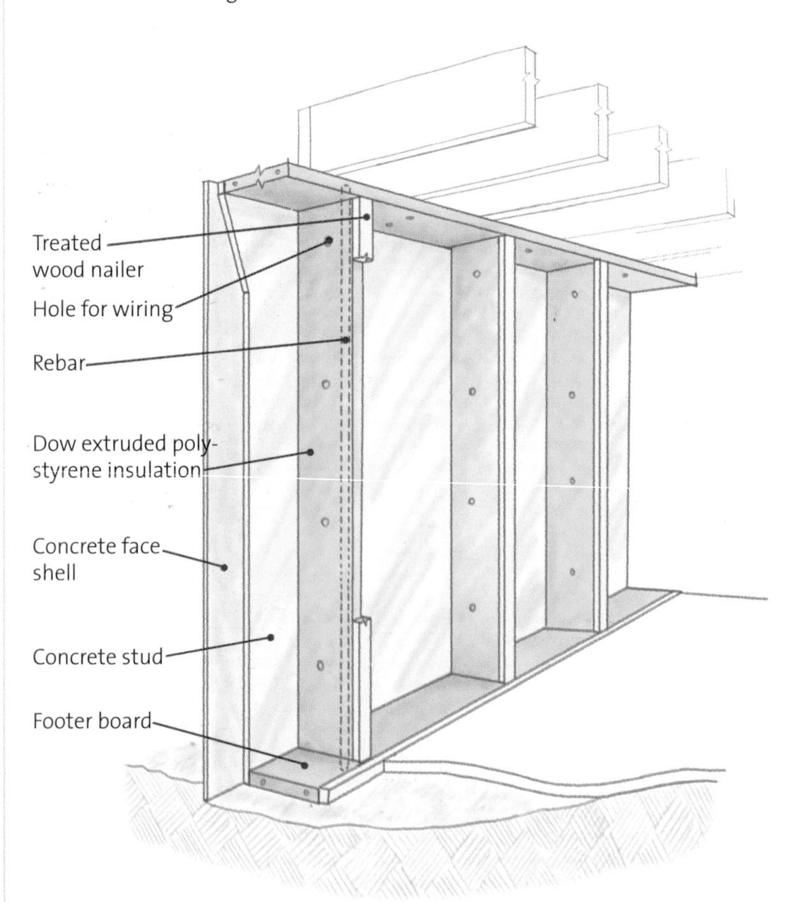

The NAHB Research Center used precast concrete basement panels for a demonstration house in Bowie, Md.

Block foundations are usually costly

There are other alternatives besides the foundation wall systems I've discussed. The most popular and perhaps least cost-effective is concrete block. The need for a footing, skilled masons, reinforcing steel, and solid grouted cells makes this old style of construction costly and slow. But if you're in a concrete-block market, some of the modern alternative systems, such as lightweight dry-stack block (Precast Autoclaved Aerated Concrete Units), surface-bonded block, and insulated concrete-form blocks represent viable value-engineering alternatives. Their cost advantage, if any, comes because you or your builder can do the work without hiring a mason.

All of these systems have excellent structural and insulating properties, and you may choose to use one for good reason, but I won't discuss them here because I've found they fall flat when compared—on a cost basis—to optimized concrete construction or treated wood.

You'll find the exception in areas like Florida and South Carolina, where above-grade masonry walls have become the norm. These systems provide viable above-grade alternatives

Insulated Concrete Forms: Warm but Costly

Because of their novelty and relative ease of installation, insulated concrete forms (ICFs) are appealing to many builders. The most popular type of ICF used in residential construction resembles large, stackable Lego blocks, which gives these forms immediate appeal for their light weight and quick, snap-together installation. Made from extruded polystyrene, these insulating forms stay in place after you fill their cavities with 3,000 to 3,500 psi concrete.

These forms provide an insulating factor ranging from R-18 to R-30. When you consider that uninsulated basement walls account for approximately 22% of a home's heating energy loss, the appeal of ICFs becomes apparent. But this system, which requires an 18-in. concrete footing, steel reinforcement, and expensive, proprietary blocks, does not provide a cost-effective alternative.

From a cost perspective, the only advantage of stackable ICFs comes with their ease of installation, which allows carpenters to place the forms and eliminates the need for a concrete-form subcontractor. If you're going to use them, consider pouring fiber-and-steel-reinforced concrete, which will eliminate most of the reinforcement bar required.

Bill Eich of Bill Eich Construction in Spirit Lake, Iowa, likes the superinsulating qualities of ICF blocks. But he readily concedes the system does not provide a cost savings over more conventional foundations.

Because of deepened frost footings that make forming difficult, in cold-weather areas, it's common to see contractors pour slabs in a costly, three-step process involving footings, a row of concrete block, and then the slab.

because they install more easily than block and provide insulation and finish options that offset some of their additional costs. Because their use comes in above-grade applications, you'll find some of these systems discussed in Chapter 4 under "Alternatives to Wood Walls."

Slab-on-Grade Foundations Are Affordable

If the primary consideration is affordability, a slab-on-grade foundation is generally the best alternative. The only exception involves sloping sites, where a crawl-space foundation works better because it costs less to build a stepped foundation wall and floor out of cripple studs and joists than it does to erect concrete retaining walls and build up a subgrade suitable for a slab. Poorly compacted fill and bad soils also can make slab construction impractical, but even on sites with expansive or collapsible soils there are alternatives, such as post-tensioned slabs. Frost considerations used to preclude slab foundations in northern regions, but thanks to the development of shallow, frost-protected footings, slabs-on-grade are now possible from Florida to Alaska. In short, there's no excuse for disregarding this option.

Monolithic slabs combine footings, walls, and floors

A monolithic concrete slab combines footings, walls, and floor into a single operation and provides the least costly foundation system available. This system is very common in warm-weather areas, but not in colder climates where contractors have gotten used to an inefficient, multistep process involving footings, walls, and slab as three separate pours.

Monolithic slab-on-grade starts with a staked layout and trenching for footings. Then, using tall stakes and braces, the builder frames one-sided forms that create the edges of the floor and foundation wall simultaneously. Once these forms are set, the plumber can

Post Tensioning for Added Strength

Pier and curtain-wall foundations are expensive options for building sites with expansive soil conditions. Instead, build a reinforced or post-tensioned slab. In place of welded-wire mesh or conventional rebar, special steel "tendons" are cast into the middle of the slab, running in both directions 4 ft. to 8 ft. on center. After the slab has cured, workers use a stressing jack to pull on one end of the tensioning rod and tighten it to an engineered load.

Post tensioning reduces deflection and cracking when the slab is placed on unstable soil. It also allows longer spans, thinner slabs, and lighter structures. Post tensioning is commonplace in many southern and western states, but it can be used wherever soil conditions make it practical and cost-effective. The technique also can be combined with frost-protected shallow footings.

Cable cast into a slab foundation is put under tension after the concrete has cured to improve the strength of the slab.

install sewer and water lines. Although footings generally take a little extra material when poured along with the slab, this one-step operation will more than compensate for an added yard or two of concrete.

Reduce slab thickness to 3½ in. or less

General practice and most codes recommend a slab 4 in. thick, but under certain conditions—requiring a carefully prepared subgrade and soils with high stability—a nonstructural slab 2½ in. thick makes an adequate and durable subfloor. The NAHB has experimented successfully with 2½-in. slabs, and they are fairly common in Europe. Most builders prefer a 3½-in. slab because variations in sub-base elevation make minimum thickness difficult to achieve reliably. But it's certainly not necessary to install an interior slab thicker than 3½ in. Residential loads are light enough that extra concrete does not provide a measurable benefit.

> If the primary consideration is affordability, a slab-on-grade foundation is generally the best alternative.

Choose concrete designed for the job

Only slabs in unheated areas exposed to freeze and thaw cycles, such as garage floors, call for high-strength, air-entrained mixes. But many builders use the same concrete mixture on all their slabs, even when interior slabs only call for a four-sack (2,500 psi) mix. Unless you're building on poor soil, problems with slab-on-grade concrete usually stem from too much water in the mix, inadequate waterproofing and drainage, and bad workmanship, not weak concrete. Using a richer mix when it's unnecessary is an expensive alternative to quality control—and not nearly as effective.

Instead of wasting money on extra cement, it's a better idea to avoid the three problems that generally cause concrete failure: adding too much water to the mix, finishing concrete before the bleed water has had a chance to evaporate, and not allowing the concrete to cure properly.

With the right preparation, footings and slab can be placed at the same time in what's called a monolithic pour.

This photo captures all the drama of a slab pour, including improperly installed welded wire mesh and a moisture barrier without a cover of coarse, dry aggregate. Both are potential problems.

Eliminate welded wire mesh

Standard welded wire mesh does not add structural capacity to the slab; it only helps control temperature and shrinking cracks—as long as it is installed correctly. Many builders don't use wire at all, and most jurisdictions don't require it. Although certain builders think that wire provides a margin of safety, unless the mesh is located uniformly in the upper third of the slab, it provides no benefit whatsoever. The only way to ensure this product works reliably is to use some form of masonry "chair," which holds the wire in place. Most builders find this process expensive and time-consuming.

Even if you go to the extra expense of using wire mesh with chairs to hold it in place, slabs greater than 10 ft. in any dimension will experience cracking due to temperature changes and shrinkage. Adding fiber reinforcement to the concrete controls cracking more effectively and saves the cost and hassles of welded wire mesh.

Some builders argue against fibers because synthetic fibers only resist normal cracking. Fiber won't prevent separations in the slab if the ground beneath it heaves or sinks (called "subsidence"). Most of us wouldn't tolerate a severely cracked slab even if strands of rusty wire held the pieces together like a broken jigsaw puzzle. It would have to be replaced anyway. As with most value-engineering techniques, the cure for heaving and subsidence is preparing the sub-base properly and compacting ditches so they won't collapse. If a concrete slab crosses a soft ditch, or needs strengthening under severe soil conditions, rebar is a better choice. It's not much more expensive than wire and a lot cheaper when you consider that rebar actually works.

> Even in colder climates, where deepened frost footings are required, a slab-on-grade may still be the most affordable foundation.

Manage cracks with control and isolation joints

The purpose of control and isolation joints is to manage cracking. After a slab has cured for a day, chalklines can be snapped at regular intervals and in places prone to cracking (such as post locations and corners). These lines are scored with a masonry blade. This approach is generally less labor intensive and more effective than tooling construction joints as the slab is finished. The recommended distance between either saw kerf or tooled joints is about 25 ft. Many builders prefer spacing between 12 ft. and 18 ft.

Isolation joints serve an entirely different purpose, which is often misunderstood. Many builders refer to the joint between a slab and an adjacent surface as an expansion or compression joint, which reveals a misinterpretation of the joint's purpose. The fiberboard or extruded polystyrene joint installed between a slab and an adjacent surface is there only to separate the materials and allow for vertical movement. Slabs don't "expand" and are almost never under compression. They do, however, lift and fall with seasonal changes in ground humidity. Slabs in unheated areas like garages are also subject to frost heave. If you use a vapor barrier, you can turn it up at the edges to form an isolation joint that lubricates the movement between adjacent surfaces without having to add a separate material.

Slabs designed for cold climates

Even in colder climates, where deepened frost footings are required, a slab-on-grade may still be the most affordable foundation. If the builder is using conventional techniques, the difference in cost between basements and slabs

TRADE SECRETS

Insulated shallow foundations provide immediate cost savings approaching 40% and, because of their enhanced insulation values, provide ongoing savings in energy expense.

BEATING THE COLD

Frost-protected shallow foundations can be used successfully in any part of the country. Heat from the building is trapped by rigid foam insulation, allowing a very shallow concrete foundation without the fear of frost heaves. The technique is slightly different for an unheated space, such as a garage. Exact specifications vary by geographic location.

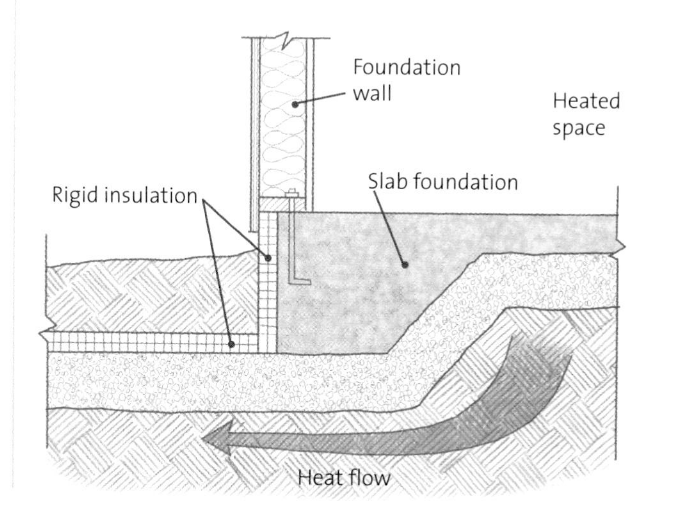

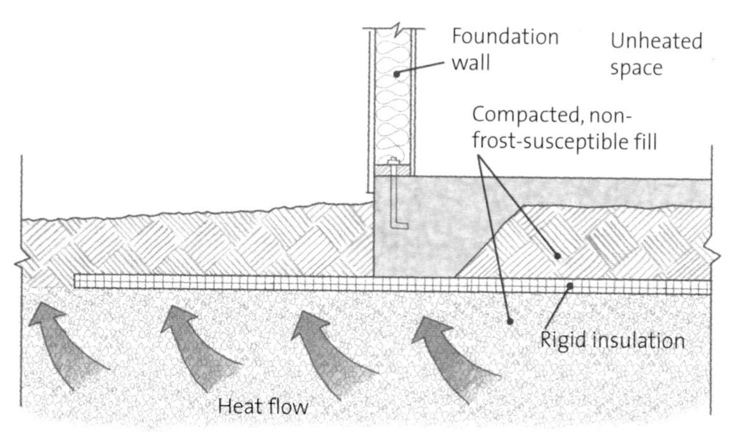

will be marginal—the builder would dig below the frost line and pour a foundation before forming the slab. Many builders disregard slabs as a foundation option because, they reason, once you've dug below the frost line, it's not that much more expensive to dig a little deeper and build a basement. This was true until codes finally began to accept the frost-protected shallow foundation (FPSF) system. It represents an alternative to deepened "frost-free" footings and can be used even in North Dakota and Alaska, saving thousands of dollars over conventional construction.

Millions of homes have been constructed economically and successfully with FPSF in Northern Europe, Canada, and the United States. Even Frank Lloyd Wright designed and built a type of FPSF intended for affordability in his Usonian homes. The principle is to magnify and preserve ground temperatures with insulation—somewhat like using mulch in a flower bed to prevent frost damage. In this case, the house acts as a warming blanket that increases the temperature of the ground around the footings. By adding insulation, FPSF technology captures this heat and magnifies it to raise the frost depth around the perimeter of the house. Even in the coldest climates, this system allows footings as shallow as 16 in.

Millions of homes have been constructed economically and successfully with FPSF in Northern Europe, Canada, and the United States.

Depending on the severity of the climate, a footing design requires between 1 in. and 2½ in. of extruded polystyrene insulation along the face of the footing and foundation wall, sometimes extending outward 12 in. to 36 in. horizontally.

Under unheated structures, additional insulation can capture and magnify ground temperatures, providing one of the most practical applications of this technology—

TRADE SECRETS

Be careful not to build any "cold bridges," which are created by excessive gaps in the foundation insulation and conduct outside temperatures into the soil beneath your footings and slab. Cold bridges increase the potential for frost heave.

namely, the elimination of expensive frost footings around the garage.

The subtleties of the FPSF system require careful study. It's important to select the right insulation, because only a few products can maintain an effective R-value throughout the expected lifespan of a building. Rigid, cellular polystyrene thermal insulation comes in two compressive strengths, either 15 or 25 psi (2,160 or 3,600 psf). Only the products that have higher compressive strength, such as DiversiFoam® Products's CertiFoam® 25 and Owens Corning Foamular 25® are suitable. Foam boards with a compressive strength of up to 40 psi are available for special situations requiring greater strength. Although not as readily available, expanded foam (bead board) in a 2-lb., 25-psi density is suitable for below-grade applications. It costs $2 to $4 less per sheet than extruded foams.

A builder contemplating the use of FPSF should know that proper specification and installation of insulation is paramount to its success. Guidelines from the IRC and the NAHB Research Center's Design Guide for Frost-Protected Shallow Foundations are useful references.

Employees from Bill Eich Construction install rigid foam insulation for a garage slab (see top photo). Over the last 15 years, Eich has installed hundreds of frost-protected shallow foundations without a single failure or foundation-related callback. Eich claims he saves $1,500 to $4,000 on every home he builds or remodels using this technology.

Sealing Foundation Walls

Leaky basement walls and slabs represent one of the major callback headaches for homebuilders. Some builders invest a small fortune in waterproofing and still have leaks. Others just roll or spray on a coat of asphalt on green concrete and never have a complaint. When a basement or slab does leak, repairs are costly. Just what type of moisture-proofing will be required depends on the site. It's worth mentioning that as much as 90% of foundation moisture problems are related to improper grading.

Damp-proofing with asphalt or plastic

Plain asphalt sprayed or rolled onto the exterior face of a freshly poured concrete wall is one of the oldest and cheapest means of damp-proofing a basement. It's also one of the least effective. Asphalt can't bridge cracks that inevitably form in basement walls, and most contractors who use asphalt won't guarantee walls against leaks. They just agree to fix any leaks during the first year after sale.

In a sense, this approach works because basement-wall repairs usually don't cost too much. Wall leaks are easy to locate, and they can be patched with hydro-cement or an epoxy injection. A sheet of polyethylene installed over basement walls acts as a capillary break. This inexpensive medium does not seal water out, but it does direct moisture down toward the gravel drain at the footing.

Dimple sheeting is another polyethylene product that provides a capillary break, but with a little more sophistication and durability than plain old 6-mil plastic. Viewed in profile, dimple sheeting looks like an egg carton with embossed bumps and dents. These dimples provide silt-free air gaps that channel groundwater toward the footing drain or gravel bed. Though more expensive than plain polyethylene (about 35¢ a square foot), dimple sheeting competes with rolled or sprayed asphalt. A crew of two people can install it over an average basement in four hours.

This polyethylene sheet resists moisture while the dimples provide air gaps and a drainage plane so groundwater trickles down to foundation drain tile.

Damp-proofing basement walls with plain asphalt is an inexpensive treatment with some limitations.

Geocomposite fabrics cost even less and provide similar protection. They consist of embossed plastic sheeting with a silt filtering fabric attached. The embossed plastic provides a drainage plane for water while the filter fabric keeps silt from clogging up the voids. Strong enough to resist backfill pressures without collapsing, geocomposites generally come in 4-ft.-wide rolls that can be nailed, glued, or taped to the foundation wall.

Using clay as a waterproofer

The products mentioned so far only damp-proof your basement walls, and this is generally enough. If you want to make basement walls waterproof, you'll have to use a product that actually prevents the transmission of moisture

TRADE SECRETS

Be careful not to overcompact the soil around basement walls, especially if the site has cohesive or clayey soils. Tamping of soil, if done at all, should be just sufficient to prevent excessive settlement of backfill. Any more than that, and you may create excessive lateral pressure.

even under severe hydrostatic pressure. Instead of buying an expensive, proprietary waterproofing system, try bentonite clay, one of the oldest and most effective methods known.

Mined in the Black Hills, this natural mineral product swells up to 15 times its dry volume when wet to form an impermeable gel. It can be applied to fresh concrete immediately before backfilling. Bentonite™ can be rolled or sprayed, or it can be purchased already laminated onto 4-ft. by 4-ft. cardboard panels that are simply tacked onto foundation walls before backfilling. Moisture in the ground dissolves the cardboard and leaves a continuous, everlasting bentonite film to protect the basement. At 60¢ per square foot, this material should be used only where excessive moisture requires it. Even at this price, bentonite costs less than many of the elastomeric membranes on the market that run from $1.25 to more than $2 per square foot and don't work any better.

TRADE SECRETS

If you have a leaky foundation wall and don't want to dig up the yard to fix it, consider injecting bentonite clay into the soil near the crack. Many mud-jacking contractors can provide this service, which generally costs a lot less than other repair alternatives.

Vapor retarders under the slab

Where floor coverings aren't affected by moisture, a vapor barrier under the slab is unnecessary. However, previously dry areas can experience problems with moisture due to increased landscape watering or temporary fluctuations in the water table. Waterproofing a slab after the fact involves expensive gel injections and seal coats, so it's better to play it safe and install an inexpensive sheet of polyethylene with seams lapped and sealed—unless you're absolutely sure. This is one of those occasions when following the adage "better safe than sorry" can save you money in the end.

Where floor coverings aren't affected by moisture, a vapor barrier under the slab is unnecessary.

A polyethylene vapor barrier should be covered with a minimum of 2 in. of compactable granular fill, such as crushed rock and sand. Concrete placed directly over plastic is susceptible to crusting, cracking, and curling. Also, keep the water-to-cement ratio of the concrete in the slab low because bleed water trapped below a finished surface can cause delaminating and blistering.

Since vapor barriers can't stop residual slab moisture from damaging floor coverings, use high-quality concrete. Concrete blended with water-reducing admixtures, cementitious fly ash, or slag can help ensure both workability and low permeability.

A CHEAPER EXPANSION JOINT

Because concrete slabs don't really expand, a polyethylene vapor barrier turned up at the edge can be used as an isolation joint rather than more expensive alternatives. It will allow movement between the slab and the basement wall.

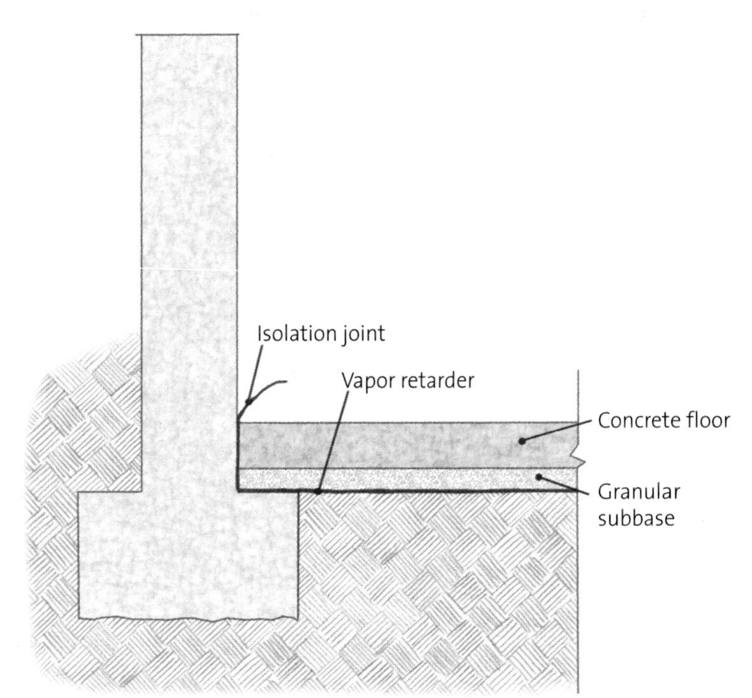

Drains are the best defense against moisture

Groundwater under the floor or foundation can wick upward through the soil and cause dampness in basements, crawl spaces, and slabs. Damp-proofing basement walls and slabs-on-grade provides a final line of defense against moisture infiltration, but the first and most important component in any foundation waterproofing system consists of proper drainage. This is one of the easiest and least expensive ways of preventing problems. Retrofitting a foundation drainage system is one of the most expensive and, unfortunately, common callback repairs in the industry.

The easiest way to ensure that a foundation drains properly is to install a 4-in. layer of crushed stone over the base of the excavated basement or slab. If gravel footings are used, just thicken the area of stone where foundation walls sit. For a traditional concrete footing, form footings over the gravel, or pipe through the footing with drain tile.

Whenever possible, provide a passive gravity-fed drainage system that takes water away from the foundation without a mechanical pump. When site conditions don't allow for a gravity-drainage system, most builders install a sump in the basement and then plumb a pump to remove any water that accumulates in the reservoir. Another approach that works well and costs less is to drop 8-in. plastic pipe into a low point along the exterior gravel bed and cut it off above grade. If necessary, a permanent sump pump can be dropped into this sleeve with a length of 1-in. PVC as a drain line. Although it's outdoors, the pump won't freeze if it is placed below the frost line.

Rectangular plastic drainage systems go in easily and quickly, providing a form and drain tile simultaneously.

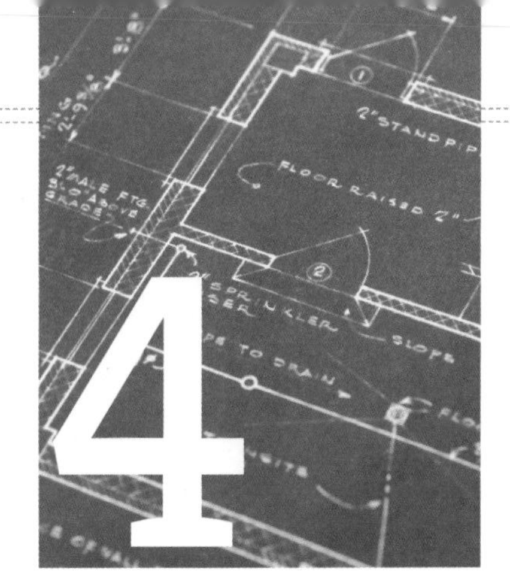

4 Framing

The abundance of natural forests in North America combined with the traditional timber skills of English settlers made wood the foremost building material in the United States and Canada. Throughout the rest of the world, masonry and concrete have eventually superceded wood. But wood remains a quintessentially North American choice for construction. More than 90% of our homes are still framed with it.

Today, a dwindling supply of old-growth timber and increased costs have forced manufacturers to find more efficient ways of using forest products. Builders have developed streamlined construction systems that incorporate engineered-wood components and framing methods that use less lumber. Research has provided builders with improved value-engineering alternatives,

Tradition and abundant natural resources have made wood the dominant building material in residential construction, but new materials and framing techniques offer many opportunities for using less of it and saving money.

In response to environmental concerns and a shortage of available timber, lumber companies have increasingly turned to engineered-wood products to replace large dimensional lumber.

TRADE SECRETS

Interest in affordable construction peaks every time the national economy falters. But the astute builder realizes that it's never wise to leave money on the table, even when business is brisk and lumber is affordable.

but builders tend to regard unfamiliar approaches with reserve. Most don't bother to optimize their framing methods, but those who do can develop a unique competitive advantage.

The wood frame represents the greatest single area for potential cost reduction. On average, advanced framing techniques can reduce the standard labor and materials bill by about 33%. In one year, I managed to reduce my framing labor costs from $5,500 to $3,200 and materials from $13,000 to slightly more than $10,000 per house.

On average, advanced framing techniques can reduce the standard labor and materials bill by about 33%.

Optimum Value Engineered Framing

To stimulate the creation of more affordable housing during the 1970s, the NAHB Research Center and HUD collaborated in a study of ways to optimize wood-frame construction. With a team of engineers and cooperating builders, they reviewed all of the value-engineering options developed to date, then tested the suitability of streamlined framing methods, such as 24-in. on-center layouts. They published their findings, built test projects, and provided builders with engineering tables and detailed instructions for a comprehensive approach to optimize wood construction.

The method promised builders extraordinary cost reductions, but only a few companies invested in the engineering and training necessary to adopt it. Even though these companies reduced their costs by about one-third, many builders hesitated to adopt advanced material-saving techniques because they feared their competitors would label them "cheap." Now all of this has changed.

Interest in Optimum Value Engineered (OVE) framing revived during the late 1990s, as homebuilders and environmentalists began to find common ground. Today, a builder can profit from advertising his or her product as

Optimum Value Engineered (or advanced) framing calls for 24-in. on-center spacing of studs, floor, and roof members.

Framing Lumber Costs

Cost of lumber in a typical 2,000-sq.-ft. house, adapted from the National Association of Home Builders. Table based on average cost per 1,000 board feet of framing lumber and 1,000 board feet of structural panels.

Cost per 1,000 board ft.	Framing lumber	Structural panel	Lumber cost per house
$250	$4,360	$1,743	$6,103
$300	$5,232	$2,091	$7,323
$350	$6,104	$2,440	$8,554
$400	$6,976	$2,778	$9,754
$450	$7,848	$3,137	$10,985
$500	$8,720	$3,486	$12,206

environmentally sensitive because one of the foremost green building techniques involves optimized wood construction. In other words, nobody will call you "cheap" for using these time-tested techniques. Instead, they'll say you're "green."

From a structural vantage point, stud layouts could actually span any number of lengths. But the 2-ft. module central to OVE works best with sheathing, drywall, and other standard materials and provides a convenient working dimension. It offers equal quality with less material and provides more space for insulation.

EQUAL VALUE, LOWER COST

Some of the basics of advanced framing include 24-in. on-center framing, an optional single top plate, headers calculated for actual loads, two-stud corners, and the elimination of jack studs and cripples over nonbearing openings. The approach is designed to save time and money over conventional building techniques.

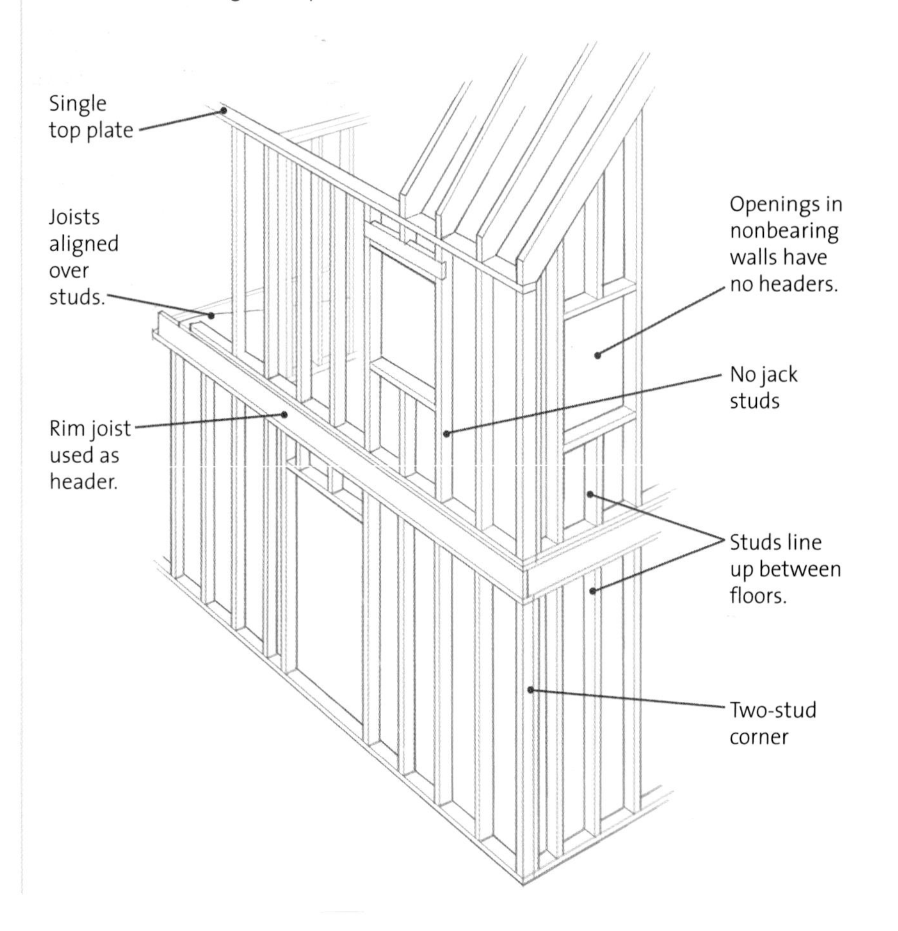

More Than a Sum of Parts

Every component in a house, from siding to drywall, makes the structure stronger. In fact, nonstructural components can account for more than 50% of its lateral resistance. The difficult task of engineering an intricate, multilayered system like a house makes it easier for engineers to rely on a simplified, component-based approach that underestimates the performance of components that work together.

This traditional "segmented" approach to engineering inevitably leads to overbuilding. A new alternative called the "whole-building" approach to residential engineering uses data gathered through extensive testing of completed structures. It attempts to bridge the gap between the simplified, segmented world of theoretical engineering and the actual strength of a whole structure. Find an engineer who is familiar with this holistic approach, and you'll avoid many of the structural inefficiencies that typically overburden homebuilders in earthquake- and hurricane-prone regions.

Saving Money with Optimum Value Engineering

Average savings obtained through the practice of advanced framing techniques, adapted from Efficient Wood Use in Residential Construction by Ann Edminster and Sami Yassa (Natural Resources Defense Council, Inc., 1998):

OVE technique	Average lumber savings per house
Size headers for actual load	$75 to $200
Design floor plan on 24-in. module	$200
Stud spacing at 24 in. on center	$200
Floor joist spacing at 24 in. on center	$400 to $700
Design roof on 24-in. module	$50 to $200

Know your material

Framing lumber comes from a variety of softwood species. Major groups of lumber include Douglas fir/larch, southern yellow pine, hem/fir and spruce/pine/fir. Each has its own structural characteristics and price.

Although your lumberyard may carry only one or two species, it pays to study span tables and determine if portions of the house can be built with less expensive lumber. While standard and better Douglas fir costs considerably more than southern yellow pine, southern yellow pine has comparable or better structural values. For example, 2x8 structural select Douglas fir joists at 24 in. on center can span 11 ft. (at a design dead load of 10 lb. per square foot and a live load of 40 lb. per square foot), while structural select southern yellow pine can span 11 ft. 5 in. However, southern yellow pine costs at least $100 less per 1,000 board feet.

Even if your structural design specifies an expensive species, such as Douglas fir, you can always substitute a less expensive material for interior nonbearing walls.

Before settling for whatever the lumberyard ships, check locally available varieties, consult span tables in the local building code, and then opt for the least costly lumber that meets your structural requirements. Even if your structural design specifies an expensive species, such as Douglas fir, you can always substitute a less expensive material for interior nonbearing walls. And you can ask your engineer to specify less expensive lumber wherever practical.

Building Walls the New Way

Since the 1970s, advanced framing techniques have established 24-in. on-center stud spacing as the norm in walls. For those of you who think this means compromised structural capacity, just know that the cherished 16-in. module didn't derive from careful engineering. It was established in England about 400 years ago—the maximum span that supported wooden lath to provide a smooth surface for the application of plaster. The advent of drywall changed everything, and unless you're building with wood lath, there's no reason to build your walls on a 16-in. module.

Although some builders cling to traditions like the 16-in. on-center convention for framing, some rules of thumb have nothing to do with engineering calculations and are instead wasteful of time and materials.

Nor do other building traditions always make sense. Break into the wall of an old house and you'll discover the original builders didn't use headers. The house still stands. Modern engineers have begun to question standards that waste one house's worth of lumber for every 20 houses built.

> Modern engineers have begun to question standards that waste one house's worth of lumber for every 20 houses built.

Economy starts with intelligent planning

The key to economical wall construction is avoiding the unnecessary use of materials and labor by planning carefully. I draw a framing elevation for every wall in my plan and study it to remove excess studs, cripples, and headers. Sometimes I relocate a window or door to make it coincide with my 24-in. modular layout. Carpenters often add studs where walls intersect, install fire blocks where they're not needed, and add double trimmers at door and window openings—traditions die hard. You'll have to work closely with your carpenters and help them understand the system.

The first rule of advanced framing is to place structural members where they belong and omit them where they are not needed. To apply this rule, you need to distinguish between load-bearing walls—which support the weight of the roof, floor joists, beams, and walls above—and non-load-bearing walls that act as partitions and support mainly their own weight.

Not all exterior walls are bearing walls, but they must be able to resist the stress of wind and seismic loads. For example, an exterior gable-end wall generally does not support the weight of a roof or floor, but it still must withstand the axial loading of wind and potential earth movement. This wall may require the shear strength provided by sheathing and hold-downs, but not the weight-bearing capacity of structural headers over doors and windows. Using a structural truss (rather than a gable-end truss) over a garage opening also eliminates the need for an expensive header.

Headers, columns, and stud layouts in bearing walls should reflect actual calculated loads and not tradition. Very few openings actually require double 2x12s. Nonbearing interior walls and some nonbearing exterior walls do not require headers, jacks, or cripple studs at all. Every building code permits this approach, and some, such as Oregon's, have actually started to encourage it.

Building codes steer decisions on studs

Codes generally limit the use of 2x4 studs on 24-in. spacing to support roof loads only, meaning you can use these studs on one-story homes or the second floor of a two-story.

Wall studs come in several varieties, including (from left) engineered wood, finger jointed, and sawn dimensional lumber.

Where 16-in. Spacing Came From

In old England, wooden lath was hand-split from logs about 4 ft. long. This thin strip of lumber couldn't hold wet plaster over a span of 4 ft., so lathers divided the distance in two by placing a support every 24 in. But even 24 in. proved too wobbly, so they split the difference once again and placed supports on equal thirds. This worked nicely, and the tradition of framing at 16 in. on center was born.

This benchmark of building remains in place neither for structural reasons nor to accommodate modern cladding materials, such as drywall and OSB. It's a 400-year-old tradition originally designed to accommodate wood lath and three-coat plaster.

ALIGNING POINT LOADS

When structural elements don't line up, it's necessary to use large headers to distribute two-story loads. When structural elements do line up, each header only carries the weight of one story (either roof or floor).

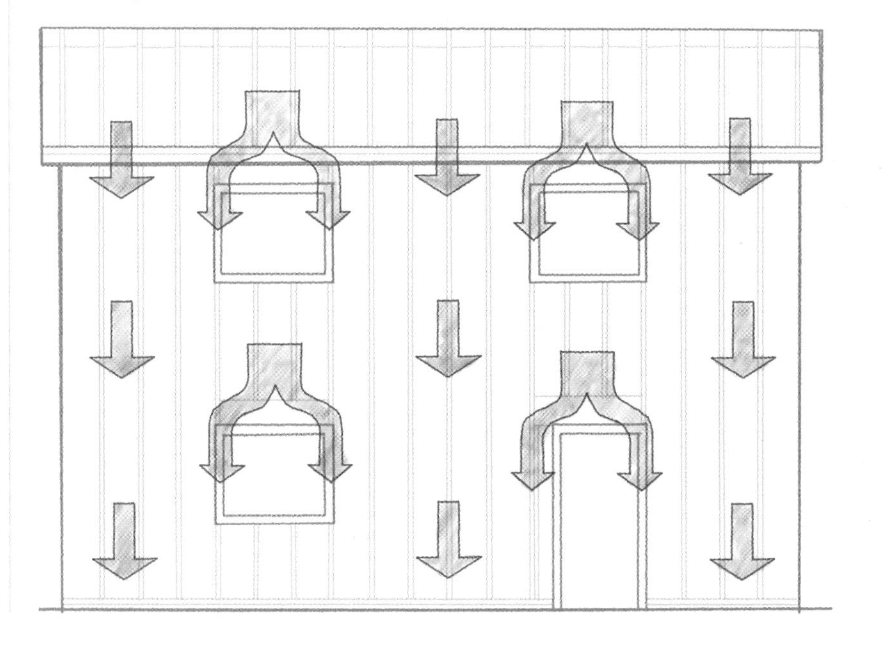

TRADE SECRETS

Reducing interior wall heights to 7 ft. 6 in. is often advertised as a way to save 5% in insulation, siding, and drywall costs. It doesn't really work. The additional costs of cutting studs, sheathing, and drywall offset the savings.

When supporting both floor and roof loads, you'll need to use 2x6 lumber at 24 in. on center. This is still cheaper than using 2x4 studs on 16-in. centers.

You can also use 4-in.-wide engineered laminated strand lumber (LSL) studs, such as Truss Joist's TimberStrand®, which are designed to carry certain floor and roof loads at 24-in. spacing. Check the manufacturer's design tables and compare the price of these engineered studs to standard 6-in. lumber.

Although prescriptive codes do not recognize the contribution of exterior sheathing and drywall, you can hire a professional engineer to review your plans and determine whether a wall assembly will meet the structural guidelines outlined in the code. Sometimes structural sheathing can add enough value to justify the use of 2x4 studs on the first floor of a two-story building.

In theory, cutting first-floor walls down to 7 ft. 6 in. (in lieu of 8 ft.) can add considerable structural value and provide some savings in exterior finishes, insulation, and paint. But in practice, I have found that the savings are negligible, while the hassles of cutting all the studs and drywall become huge. Another OVE suggestion that pays dubious job-site dividends concerns the use of a single top plate and skinny mudsills.

Stick with a double top plate

If you build with a traditional double top plate and a rim joist, it's not necessary to line up studs on first and second floors. But a favorite recommendation of advanced framing advocates is to eliminate the double top plate and replace the 2x4 bottom sill with 1-in. material. Besides costing more than a standard 2x4, the 1-in. sill does not provide much room for a nailing base. The single top plate option works, but if you decide to eliminate your second plate, you won't have the same load-spread capacity, which becomes critical to stack your framing.

Money-Saving "Green" Checklist for Wood Framing

- Use 24-in. on-center spacing.
- Right-size structural components.
- Use OSB sheathing.
- Use finger-jointed studs.
- Use factory-built wall panels and roof trusses.

The problem with eliminating the second top plate—structural considerations aside—is that it serves a practical function in lining up and connecting walls during construction. In practice, having to connect intersections and splices with metal straps takes more time and expense than simply adding the second—albeit structurally redundant—top plate. On the finish end, if you remove the second plate, the walls end up 1½ in. shorter, and many drywall contractors add an extra charge to rip down sheets of drywall. If you use 104⅝-in.-long studs to compensate for the lost plate height, the added cost more than offsets any savings.

Surveys done by the NAHB Research Center indicate that houses built with lapping top plates have much higher wind resistance.

An optimized header using only one 2x10 meets engineering requirements for this opening while allowing plenty of room for insulation.

Top plates also provide an excellent connection between perimeter exterior walls and intersecting interior walls. Although in theory this connection isn't necessary—codes don't even require it—surveys done by the NAHB Research Center indicate that houses built with lapping top plates have much higher wind resistance.

In certain areas, such as crawl-space foundation walls and the crippled foundation walls in split-level and daylight basement construction, the double top plate is indeed unjustified, and I always eliminate it.

When you frame walls within a basement, there is no concern for wind shear. If you frame with cripples over a partial basement wall, the rim joist acts as a lateral beam, and your floor joists and sheathing tie into interior walls to transfer lateral shear. Although you could say the same about the first floor of a two-story home, you certainly cannot make the argument on the second floor, where lapped plates provide a decided structural advantage.

Use two-stud corners, not three

As a holdover from timber-framed construction, carpenters used to install a 4-in. square post at every exterior corner. This led to the common practice of building three-stud corners, but neither has any structural basis. With few exceptions, the maximum load on a corner is about one-half that on a regular stud; therefore, the two-stud corner recommended by the OVE method is more than adequate. Two-stud

DRYWALL CLIPS, NOT STUDS

Two-stud interior and exterior corners require drywall clips. If the drywall contractor balks, a three-stud corner works just as well.

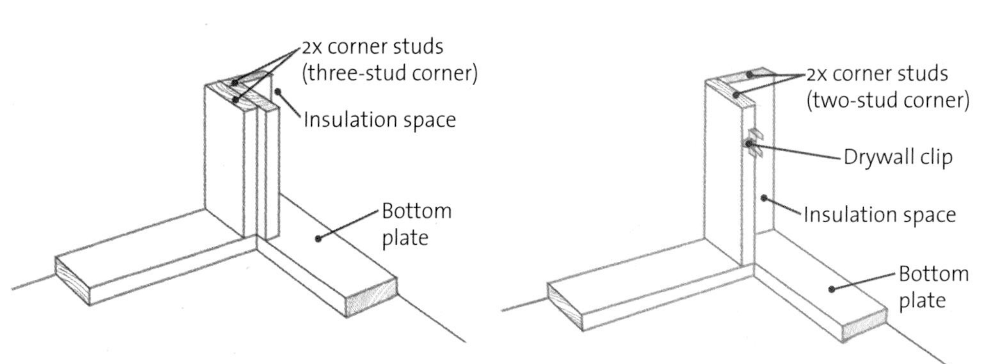

Optimized headers are engineered one opening at a time.

corners have the additional advantage of allowing room for insulation. You can also configure three studs to allow for insulation by using one stud as a cleat.

Whether interior or exterior, two-stud corners require a drywall clip, ladder blocks, or a cleat for drywall backing. You can connect intersecting walls with a single stud, avoiding the traditional three-stud partition post. This saves framing on all walls, but on exterior walls, it also facilitates insulation and avoids cold joints.

Skip unnecessary blocking

In walls that are less than 10 ft. tall, fire blocking is not necessary. In some jurisdictions, codes allow fiberglass insulation as a substitute for wood blocking along the perimeter of chases and platforms, and as midspan blocking for tall walls.

> In walls that are less than 10 ft. tall, fire blocking is not necessary.

Midspan blocking is not necessary to prevent studs from bending when they will be covered with drywall or sheathing on at least one side. If a wall remains bare, as in the center-bearing wall of a basement, then good practice dictates blocking on at least every other stud bay or, more simply, a 1x4 horizontal ledger nailed onto one side of the studs to keep them from warping.

Conventional headers waste material

The traditional method of placing double 2x12 lumber over every opening is perhaps the most wasteful of all conventional framing practices. You can save a lot of money by designing every header for its actual load. If the load is zero, then don't install a header.

On many first-floor and basement walls, the rim joist can serve as the header. At any wall running parallel to the floor joists and roof trusses, no header is typically needed. On nonbearing interior walls, a nailing frame is plenty. The same goes for jack (jamb) studs and trimmers. You can eliminate these at any opening where a structural header is not required.

ALTERNATIVES TO CONVENTIONAL HEADERS

Many builders either install headers where they aren't really needed, or overbuild them and waste materials. This alternative to a conventional header uses less lumber and allows room for insulation.

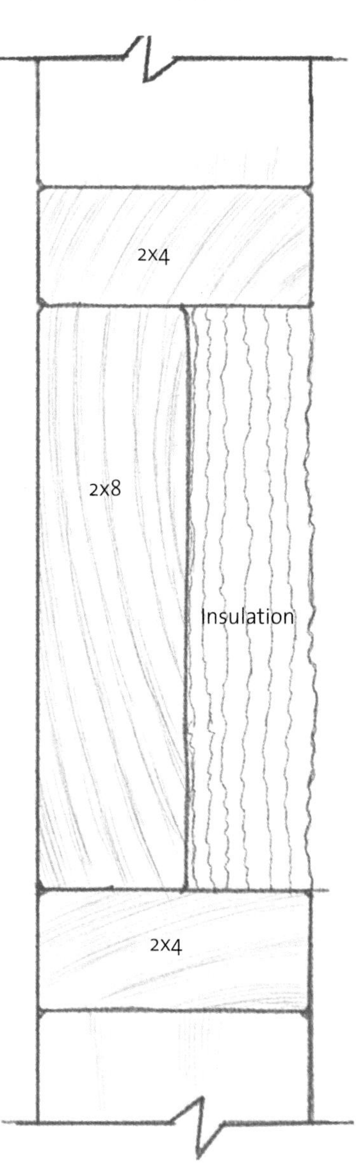

When headers are necessary, design each one to precise requirements. Then look for economical alternatives, such as engineered LSL, a glued and nailed plywood box, or single-member headers. Above large openings, such as garage doors, you can use engineered-lumber headers or even a lightweight truss. If you're careful to design the garage door and other large openings into gable-end walls parallel to any floor loads, the expense of this header can be greatly reduced.

> On many first-floor and basement walls, the rim joist can serve as the header.

OSB sheathing is a good all-around choice

As a structural element, wall sheathing joins together individual framing members to form an assembly that is stronger than the sum total of its parts. Structural sheathing eliminates the need for let-in bracing. It is rated for either 16-in. or 24-in. on-center framing (the ratings are stamped on the sheathing material). Be sure to use material rated for 24 in. on center if you're using advanced framing techniques.

For general-purpose sheathing, 7⁄16-in. OSB applied to walls vertically is economical and provides excellent backing for any type of siding. Market conditions don't always favor OSB, in which case there are alternatives: Half-inch medium-density fiberboard (MDF) provides many of the benefits of OSB, but it does not have the same structural value, which means plywood or OSB panels will be needed in discrete locations for bracing. Some panels, such as those made with MDF, provide equivalent structural support to OSB, but only at 16 in. on center. Panel prices fluctuate, but whenever the market allows, opt for OSB for its structural benefits at 24 in. on center.

Some sheathing materials, such as plastic-foam boards, provide insulation and backing but do not provide structural support. Because these products cost more than OSB and MDF, and you have to install them with supplemental bracing, they don't provide any advantage in economy construction.

There also are structural siding panels that combine the function of siding, sheathing, and bracing into a single layer, such as Texture 1-11. These products represent the most economical sheathing available. Some structural siding panels come preprimed.

In areas where sheathing is not required, many builders apply lap siding directly to studs. These builders usually opt for 16-in. on-center framing with traditional let-in or metal T-braces and/or foil-faced hardboard panels. Unfortunately, in spite of tighter stud spacing, lap siding applied directly to framing telegraphs every wall imperfection.

Some regions require shear walls

In areas where sheathing must resist the powerful lateral forces of earthquakes and hurricanes, it becomes part of a highly engineered and costly assembly called a shear wall. One reason shear-wall construction is so very expensive is that traditional engineering considers each

TRADE SECRETS

When panel prices climb and OSB is no longer the best alternative, consider switching your stud spacing on exterior walls back to 16 in. on center and use medium-density fiberboard, which provides equivalent structural support to OSB with studs spaced at 24 in. on center. Continue using the 24-in. spacing for all your interior walls.

You have many choices for sheathing, including Blackjack™ (the most common insulating board), plywood, and OSB. From a cost-to-benefit ratio, OSB provides the best price to structural benefits.

Shear-wall construction adds thousands of dollars in cost to homes built in high-wind and seismic zones.

segment of a wall separately—as if each elevation were a collection of narrow columns. Columns have limited lateral strength and require multiple hold-downs. This method of shear-wall design—known as the segmented shear-wall design approach—represents standard practice. It remains regrettably conservative and leads to the costly overconstruction of many homes or parts of homes.

Perforated Shear Walls A new method of shear-wall design known as the Perforated Shear Wall (PSW) design is more efficient and less expensive for many applications where the beefy segmented shear wall is not necessary. The American Forest and Paper Association, the NAHB Research Center, and others have been working on validating the system, pioneered more than 20 years ago in Japan, for use in the United States. It has been gaining popularity and code acceptance in coastal areas where hurricanes and earthquakes generally result in greater demand for engineered shear walls.

Prefabricated shear-wall assemblies, available in either light-gauge steel or wood-panel versions, are easier to install than site-built versions and offer greater design flexibility.

SIMPLER SHEAR WALL DESIGN

A Perforated Shear Wall (PSW) is fully sheathed with structural panels (plywood or OSB) and has openings, or "perforations," for windows and doors. The technique simplifies construction by providing a design to the entire structure as one unit, in which all the structural elements work together to distribute shear forces. It reduces costs by eliminating redundant hardware, blocks, and shear nailing.

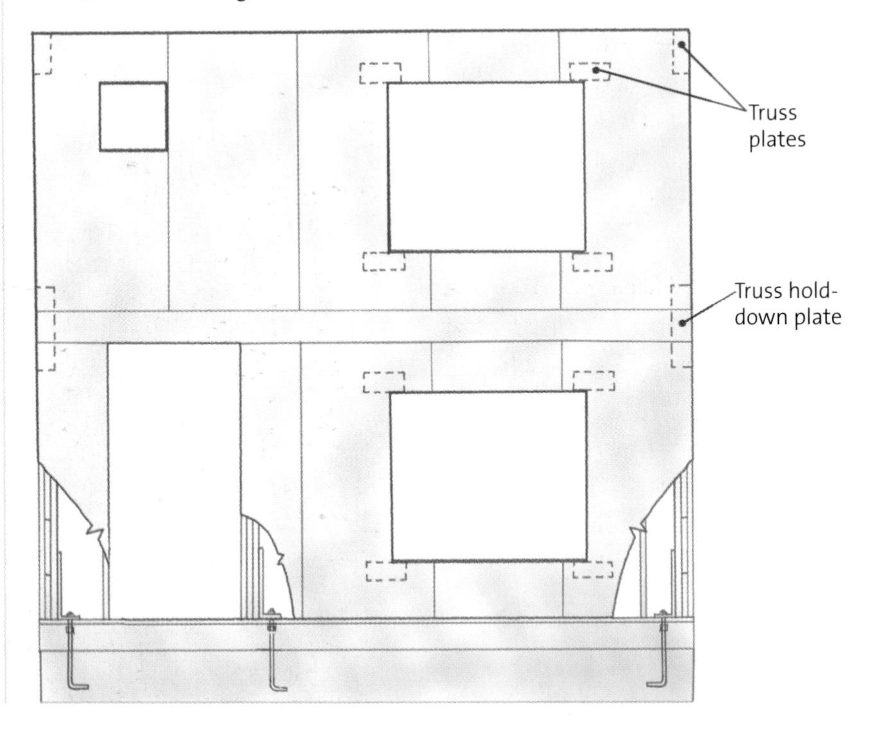

Like other progressive engineering systems, PSW takes a holistic approach to shear-wall design. Instead of regarding each elevation of a house as a collection of separate structural elements, PSW attempts to address the total structural contribution of every square foot of wall. The result is a shear-wall design that meets code requirements with fewer hold-downs. To achieve higher values when necessary, engineers enhance the shear-wall nailing schedule instead of adding expensive hardware. In one project I analyzed, we saved approximately one-half the conventional hold-downs for a savings of nearly $2,500 on a single-story ranch.

> When comparing the cost of shear-wall segments, it's important to take into account shear value and not wall size.

Prefabricated systems save time on site

Shear-wall stiffness decreases along with the ratio of wall length to height. As a result, shear-wall design requires an inordinate amount of hold-down hardware, sheathing, and strapping to build narrow panels, such as the columns typically found flanking garage doors. Several companies have developed prefabricated narrow shear walls in either light-gauge steel or wood-panel construction.

These panels are easier to install than site-built shear walls and provide greater design flexibility, including walls as narrow as 12 in. The Hardy Frame® light-gauge steel-frame system has gained popularity among Southern California framers for its ease of installation and lower cost. When comparing the cost of shear-wall segments, it's important to take into account shear value and not wall size. Increased value can mean using a smaller segment or fewer panels. On the other hand, a structure with one side completely open is often structurally stable due to the strength of the remaining three sides. Therefore, with proper engineering, such frames may not really be necessary. Regulatory requirements that provide some rather arbitrary restrictions to efficient design (particularly in high seismic areas) may make this approach costly in terms of local approval, even with the best of engineering.

Alternatives for building partition walls

Openings in partition walls don't require headers or trimmer studs. All you need is a nailing surface for trim. Partition walls can be built with utility-grade studs and, where it won't interfere with trim, 3-in. framing material (although this usually costs more than standard 4-in. studs). In certain circumstances, such as closet walls, even 2x2 studs or 2x4s turned flatwise suffice.

To minimize closet-door framing, use full-width and full-height sliding doors. This method requires no jambs or bulkhead while providing occupants with a large opening and ready access to shelves.

When anchoring a nonstructural partition wall between floor joists or truss chords, use a block ladder at 24 in. on center to secure the partition and provide drywall backing. If you're using ⅝-in. or thicker floor sheathing, there's no need to provide any special framing under partition walls.

> Openings in partition walls don't require headers or trimmer studs. All you need is a nailing surface for trim.

Metal framing has some advantages

Many builders have discovered light-gauge steel framing for interior partitions. Although steel has not proven cost-effective in whole-house construction, its light weight and workability make it attractive for some applications. Along the hurricane coast in Florida, where builders use masonry or concrete exterior walls, steel has become the material of choice for interior construction. Likewise, in areas of termite infestation, steel provides a good alternative.

For builders in the rest of the country, light-gauge steel studs represent an excellent cost-saving product for the interior of shell homes (where the exterior walls hold up the roof and floors without interior bearing walls) and for framing in basements. I have found that steel framing is excellent for bulkheads and furring walls—especially when drywall crews can complete the work of building chases and dropped ceilings for the mechanical trades without having to call back the framers.

A Hardy Frame hold-down template and wall section, popular among California framers.

Fire walls

As housing density increases, fire walls are becoming an increasingly common building requirement. For the wall to be acceptable, a recognized testing laboratory must rate it under established standards, so it's hard to get too creative with a fire wall. Yet many approved assemblies exist, so you can obtain a list of proven alternatives from either the Gypsum Association or the National Concrete and Masonry Association.

Duplex fire walls Duplex construction and detached zero-lot-line homes require a one-hour fire-wall separation similar to that required for an attached garage. You can achieve this separation easily by laminating both sides of the occupancy wall with ⅝-in. type X gypsum wallboard from the foundation wall right up to the roof sheathing, or by creating a one-hour envelope with ⅝-in. type X drywall on the occupancy wall and ceiling.

Two carpenters for Mercedes Homes in Melbourne, Fla., complete the interior of this 1,400-sq.-ft. home in two days using light-gauge metal frame partition walls.

The least expensive firewall alternative is a wood-framed gypsum-board sandwich with either one or two layers of ⅝-in. type X drywall from foundation to roof sheathing with joints taped.

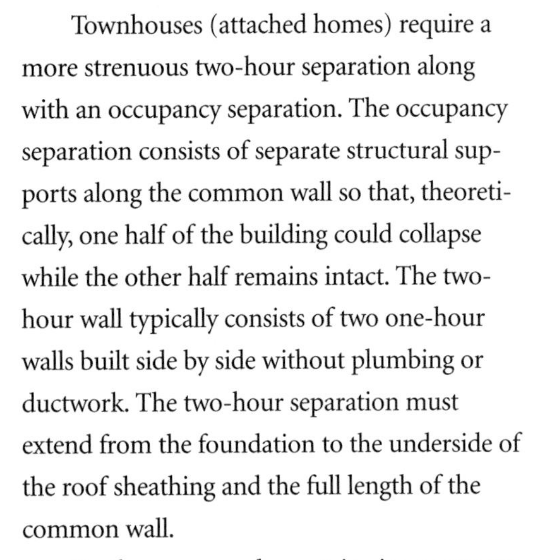

Townhouses (attached homes) require a more strenuous two-hour separation along with an occupancy separation. The occupancy separation consists of separate structural supports along the common wall so that, theoretically, one half of the building could collapse while the other half remains intact. The two-hour wall typically consists of two one-hour walls built side by side without plumbing or ductwork. The two-hour separation must extend from the foundation to the underside of the roof sheathing and the full length of the common wall.

Furthermore, under certain circumstances, you're required to build parapets and install fire-resistant sheathing. These requirements make townhouse construction inherently expensive. Don't make the mistake of assuming that the savings on one elevation of siding and windows somehow reduces the price of attached house construction over detached, single-family dwellings. The cost savings comes from fitting more houses onto a block of land. Attached housing only serves to increase density, not necessarily decrease costs.

SPEEDY CHASES AND BULKHEADS

Building chases, bulkheads, and dropped ceilings from steel studs can save time and money if you do it like commercial drywallers do—make drywall part of the structure. Many builders frame with metal studs—one stick at a time—just as they would with lumber before skinning the structure with drywall. This defeats the cost advantage of light-gauge steel framing.

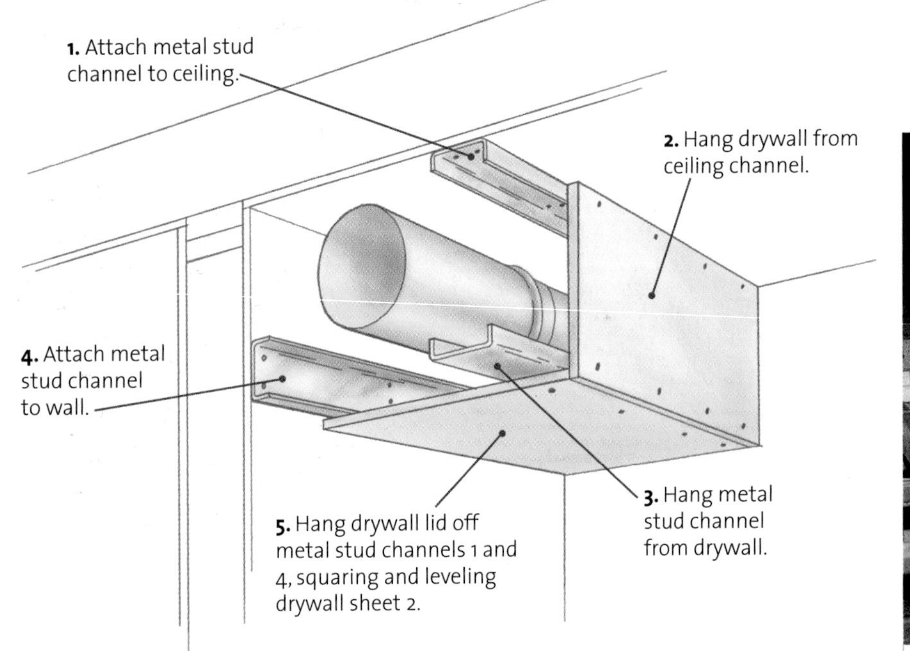

At the Millard wall plant in Lincoln, Neb., workers factory-frame about 12 homes every day.

Prefabrication Is the Future

Just as manufactured roof trusses have overtaken traditional roof framing, wall panelizers and the timber industry at large predict that manufactured wall sections will overtake site-built stick homes within a decade. Having used factory-framed wall panels for several years, I tend to agree with this prediction. This is especially true in earthquake-prone areas, such as California, where wall-panel manufacturers turn out completely engineered framing packages built with computerized accuracy.

Many framers are reluctant to try the panelized system but become panel advocates when they do. A factory charges a lot less for framing walls than a carpenter can to build them on site. This usually translates into a slightly higher profit for the framing contractor and a substantial price reduction for the builder. For the builder, quality, consistency, and control provide an even greater advantage than simple cost reduction. With factory-built walls, it's easy to apply all of the advanced framing techniques discussed in this chapter. Details are easy to specify, including cabinet blocks, tub and bathroom-accessory backing, electrical chases, and heat returns. You can design an optimum header for every opening and eliminate unnecessary cripples and studs.

Once the walls are almost perfect, a builder can walk the job with a clipboard in hand and make changes that are even more fastidious—eliminating even a single stud or block, or moving a window over a few inches, can save money next time. Once the panel-plant engineer inputs these details into the factory's computer, the changes will follow every job thereafter.

For builders, purchasing preframed components also provides a more rational cost basis for change orders than negotiating with your framer. The panel plant charges for actual labor and material instead of using round numbers pulled from the sky. For example, adding a

Factory-framed walls arrive at the job site before the concrete contractors have pulled off their equipment.

One advantage of using factory-made wall panels is very precise engineering. Careful inspection of components leads to the elimination of unnecessary building components and lower building costs.

Concrete block is the most expensive and least efficient alternative to conventional wood construction. It is used in about 5% of the nation's houses.

Many alternative products, like these concrete insulated form blocks, have initial appeal because of their apartment simplicity of installation and energy benefits, but none have proven cost-effective.

closet to one of my plans cost me $13.82 for extra labor. Builders who shift more of their expenses from subcontractors to suppliers can get better payment terms and lower insurance premiums to boot.

Alternatives to Wood Walls

Builders in the United States construct most houses with wood for good reason: There is no cost-effective alternative. Although there are promising technologies, some of which provide specific benefits, not one can rival wood for availability, ease of construction, and bottom-line cost.

Among the alternatives are concrete block, steel framing, insulated forms, foam-core structural panels, autoclaved cellular concrete, old-fashioned cob, and even straw bale. Surprisingly, the most inefficient and expensive of all these systems, concrete block, represents about 5% of our nation's homes. All of the other systems combined account for less than 1%. The principal factors holding back the alternatives are the lack of a skilled workforce and proprietary technology that adds a premium that most production builders are not willing to absorb. If you are building your own home and labor is free, an inexpensive resource such as mud for cob or straw for straw-bale walls can provide a reasonable alternative. But on a commercial level, these systems are not ready for production.

If you are building your own home and labor is free, an inexpensive resource such as mud for cob or straw for straw-bale walls can provide a reasonable alternative.

To test this somewhat oppositionist opinion, I called Nevil Eastwood, the national construction manager for Habitat for Humanity℠, the largest affordable homebuilder in the United States. Having overseen the construction of thousands of affordable homes employing every type of construction system, from standard to cutting-edge, Eastwood echoed my conclusion that, for most projects, nothing works as well as wood.

The exception comes when environmental or market conditions dictate different priorities. For example, in colder climates the high insulating values and easy assembly of foam-core structural insulated panels offset their cost and provide a viable alternative. In other areas, high winds and heavy termite infestation dictate the use of masonry and steel as preferred building materials. In these areas, production builders are experimenting with new technologies to speed up construction and make homes more affordable to build.

Floor Framing

A wood floor combines joists, girders, and sheathing to support the building's live and dead loads. Most wood-framed floors use nominal 2-in.-thick joists spaced at 16 in. or 24 in. on center, supported by foundation walls at either extreme and a beam or wall in the middle. OSB sheathing, fastened to the joists with glue and ring shank nails, completes the assembly.

A Mercedes Homes concrete house going up in Melbourne, Fla.

To economize on materials, use a 48-in. design module perpendicular and parallel to the floor framing. This not only saves sheathing, but it ensures you won't have to trim joists at the ends and won't waste an extra joist lengthwise to stay on module. Locate bearing walls perpendicular to joists and, whenever possible, line up joists with wall studs and roof framing using a 24-in. "point load" modular framing system.

To economize on materials, use a 48-in. design module perpendicular and parallel to the floor framing.

Try to keep floor spans to a minimum. Narrower houses cost less to build because they require smaller joists. Even if a house is not narrow, minimize the horizontal distance between supporting walls to reduce the size of joists. Whenever possible, use an open floor plan on the second story and not the first. By placing bedrooms and other small rooms downstairs, you provide intermittent support under the floor, resulting in shorter joist spans and therefore smaller, less expensive joists.

Size joists by calculating loads

Joist requirements should be calculated on actual occupancy loads: 40 lb. per square foot is required for living areas but only 30 lb. per square foot for bedrooms. Choose the best species of dimensional lumber according to the load and span requirements. For example, if the joist span allows the use of hem/fir instead of Douglas fir, it will save a few hundred dollars on a single floor. Likewise, if you can use spruce/pine/fir instead of hem/fir, you can save even more.

Explore joist options

Whatever the species of lumber, it's generally less expensive to use a larger dimension joist at 24-in. on-center spacing with ¾-in. tongue-and-groove sheathing than 16-in. spacing with ⅝-in. sheathing. But you should explore alternatives, such as 32-in. joist spacing with 1-in. tongue-and-groove floor sheathing, or even 12-in. spacing with a smaller joist, such as a 2x6, along with inexpensive, ½-in. sheathing.

Saving with 24-in. Spacing

Sample cost comparison between 2x8 joists spaced at 16 in. on center and 2x10 joists at 24 in. on center for 1,600-sq.-ft. ranch plan.

Material	Spacing	Total cost
2x10	24 in.	$2,296.80
2x8	16 in.	$2,371.50

Wood I-joists are increasingly common in residential construction, making possible longer spans than conventional lumber of the same height.

Cross-bracing between floor joists is an old building tradition, but tests show it is not necessary for a floor system with joists 12 in. or less in height.

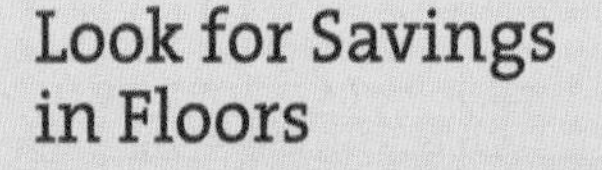

Look for Savings in Floors

More lumber goes into a floor than any other part of house except the roof. The widespread use of engineered roof-truss systems has reduced waste. Floors, on the other hand, provide a gold mine of cost-cutting opportunities.

Simply increasing joist spacing from 16 in. on center to 24 in. can shave $75 to $100 off the cost of a typical two-story house. Streamlined detailing at stairwell openings, rim joists, and wall-bearing points can help lighten the structure even further. Since floors consist of large-dimension lumber or pricey engineered joists, even minor reductions in material and labor can have an impact on the bottom line.

Most builders find that 2x10 joists run at 24-in. on center with ¾-in. tongue-and-groove sheathing represents the optimum balance between material cost and labor savings. The larger joist bays provide more room for ducts and plumbing. Although a 32-in. module provides even more room, it's more difficult to make this dimension coincide with wall and roof framing.

Cross-bracing is a wasteful practice Carpenters used to consider cross-bracing or blocking between floor joists as "good practice" to counter lateral torsion and make for an all-around stiffer floor system. But full-scale engineering tests conducted by the NAHB Research Center in 1961 proved that cross-bracing and blocking is unnecessary with dimensional floor framing up to 2x12s. Nearly 45 years later, builders continue the wasteful practice. Don't follow this tradition blindly unless there's a building official with a bayonet at your back.

Double joists sometimes not needed It's also traditional but unnecessary to install double joists under nonbearing partitions. If you use ⅝-in. or thicker sheathing, it's strong enough to carry the load of a partition wall without adding an extra joist or even blocking. But sometimes nails driven into sheathing can cause squeaks as the floor sheathing deflects with live loads while the wall remains rigid. You can avoid this problem by gluing down the sill plate when fastening it to the sheathing. Use glue whenever you place a wall over a long, unsupported span. Be aware that this problem occurs more often with engineered floor systems that allow greater span flexibility (more about this later).

Make the sill plate smaller In conventional construction, framers use a 2x6 pressure-treated sill plate to anchor the floor system to the foundation. When building on a concrete or masonry stem wall, a 2x4 plate is enough, regardless of the thickness of the foundation wall or the size of the joists. In many foundation systems, such as permanent wood and precast foundations, a sill plate is redundant. Joists can be toenailed directly to the foundation wall or anchored with straps every 48 in.

Reduce the size of the band joist The band joist (sometimes called the rim or header joist) provides a good layout ruler and a brace to line up and support joists before sheathing the floor. But if you're using the point-load framing system—in which joists, studs, and roof framing members are stacked in line with each other—there's no need to use a band joist that's the same size and quality as your floor joists.

For convenience, use strips of sheathing or a temporary nailing board to lay out the joists and provide bracing, or just tack the perimeter sheathing in place as you roll out the joists. You can run the wall sheathing down over the edges of the joists and create a box-beam effect by nailing the sheathing first to the foundation sill plate and then to the butt ends of the joists along with the sill plate for first-floor walls. If the building inspector requires additional blocking between joists, you may want to follow tradition and install a band joist for convenience.

> An extremely useful technique developed by the NAHB Research Center involves splicing dimensional joists to create single, long-span members.

When a band joist is used, there's no need to buy structural-quality lumber because the band joist is not a bearing member—any dimension board will do. The one exception comes when you have openings, such as a basement window. In this case, the band joist can stand in as an inexpensive header. If the opening exceeds 24 in., you'll have to double the band joist and possibly attach floor joists with hangers.

Use off-center spliced joists An extremely useful technique developed by the NAHB Research Center involves splicing dimensional joists to create single, long-span members. Unfortunately, builders have not adopted this system, although it affords an inexpensive alternative to engineered floor joists and provides a stiffer floor with smaller-dimension lumber and virtually no cutting.

The idea is to connect two standard-size joists of unequal length with a plywood gusset or metal truss plate. Each splice has to occur at

INCREASING ALLOWABLE JOIST SPANS

Research conducted at the NAHB Research Center developed an off-center spliced-joist design that can be fabricated on site with standard dimensional lumbers. This method lengthens the allowable span of lumber joists.

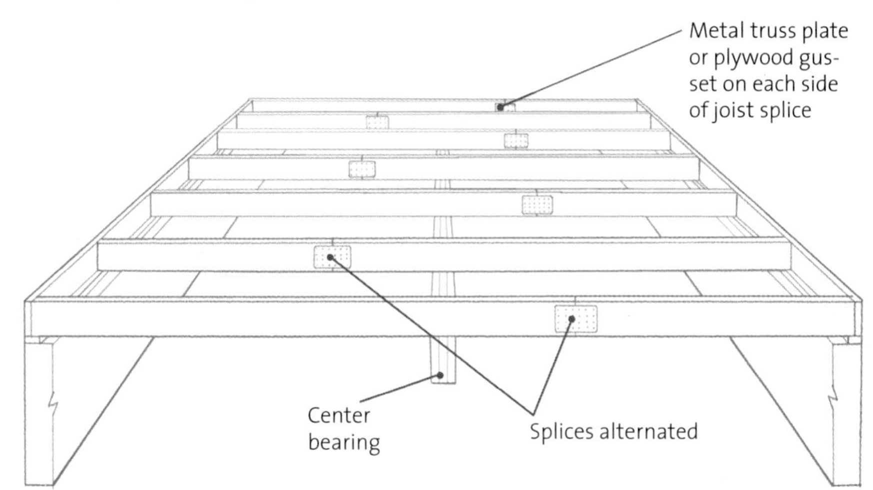

TRADE SECRETS

Proper spacing is especially important when using OSB floor sheathing because it swells when it gets wet. Some contractors think that this rule does not apply in arid areas, but it does: Indoor air can contain high levels of humidity in any climate.

TRADE SECRETS

When you need the width of two joists to anchor a railing, just add a 2x4 ledger on the floor joist instead of a more expensive full-depth board.

a point of minimum bending stress and off center. The technique works best with joist spans divided in equal spans over a center support. Adjacent joist splices are located on alternate sides of the center support. Not only can you reduce the width of the joists because the structural continuity provided by the splice creates a stiffer floor frame, but you eliminate the hassle of lapping joists and trying to install sheathing over a 1½-in. offset. In essence, you get some of the benefits of an engineered floor system without the added cost.

I have applied this technique with the help of my local truss manufacturer, who was able to deliver 2x8 truss-plate-spliced joists straight from the plant. The off-center splice system allowed me to use 2x8s at 24-in. instead of 16-in. centers. The advantage of using a truss plant instead of plywood field-built gussets comes, in part, because the plant can supply the engineering needed to convince the building department that this technique is suitable. (You can find detailed information in the NAHB Research Report No. 4, *Off Center Spliced Floor Joists* (NAHB Research Center, 1982.)

TRADE SECRETS

Although cantilevered framing provides an efficient means of adding floor space, you should avoid cantilevered balconies and decks. Waterproofing and insulating the indoor/outdoor transition adds substantial cost and is a long-term maintenance headache. Instead of cantilevering your joists, attach balconies and decks to the exterior shell.

Cantilevered floor designs You can usually cantilever floor joists up to 24 in. parallel to the length of the house to add extra floor area without enlarging the foundation. This technique is useful for creating extra floor space for pop-out closets and bathrooms. You can simply extend the roofline at a slope without any concern for lost headroom. Always cantilever parallel to the joist layout, never perpendicular to it. The extra framing required at right angles to the joists eliminates any economic advantage.

> You can usually cantilever floor joists up to 24 in. parallel to the length of the house to add extra floor area without enlarging the foundation.

Instead of fitting rim blocks between joists right over the sill plate, which separates the cantilever portion of the floor from interior framing, place blocking inside the building and within 12 in. of the rim. This allows the cantilevered floor to be insulated without creating a thermal break or interrupting framing operations to install insulation. If you toenail joists into the sill plate, you can use a 2x4 block at the top of the joist in lieu of wasting money with full-depth blocks. This also makes it easier to run wires, ducts, and insulation into the cantilevered area. To provide edge nailing for a soffit under the cantilever, run flat blocking along the edge of the mudsill.

If you're building a crawl-space foundation

ADDING A CLOSET IS EASY

By extending the roofline over a 2-ft. cantilever, it's possible to add a closet without complicating framing or making the footprint of the house any larger than it has to be.

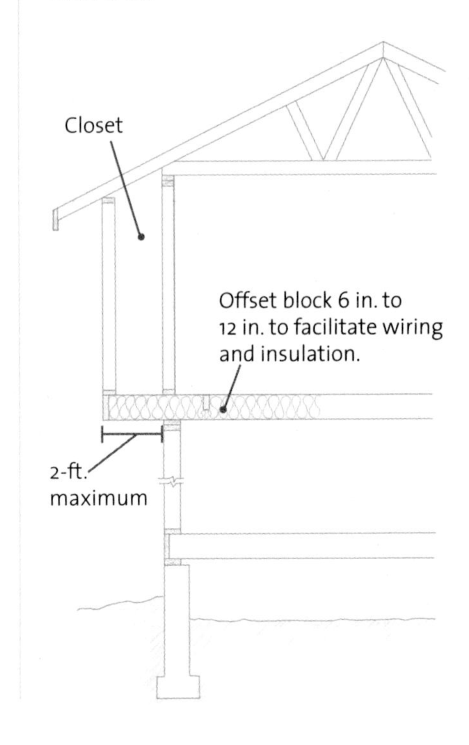

that will require more than one girder to support the joist span, you can save considerably by using an innovative technique developed in Lacey, Wash. Instead of running joists from one foundation wall to another, locate girders within 4 ft. of the exterior walls and then allow joists to cantilever 2 ft., without actually coming into contact with the foundation wall.

The sheathing can span the distance between the end of the joists and the mudsill, eliminating the band joist entirely and saving about 50 ft. of joist for every 10 ft. of floor. A builder developed this design for a demonstration project on innovative OVE construction. It saved the builder about $850 and illustrates the creative thinking required for value engineering.

Stairways

Stairwells can consume many construction dollars if you're not careful. To reduce costs, build stairs as a straight run, parallel to floor joists, and avoid winders, U-turns, and stairwell platforms. Locate the stairway opening to coincide with the 48-in. framing module and limit the width to 4 ft. This allows the use of one less trimmer and eliminates the need for double headers. If you locate one of the headers within 4 ft. of the joist ends, you don't need the extra support of any double trimmers at all, saving another joist.

When a stairwell must be located perpendicular to the floor joists, try to place at least one side of the opening within 6 ft. of a girder or bearing wall so that you can end-nail tail joists and avoid the time-consuming task of installing joist hangers. To assure a good connection, end-nail any tail joists that are 6 ft. or shorter through a single header first and then attach the second header. Of course, headers will require hangers where they attach to trimmer joists, and any tail joists longer than 6 ft. will require hangers, too.

Locate a small opening, such as a crawlspace access, within an existing 22½-in. joist

Economical Stairs

After a careful study of the table, it's clear that the optimum stair assembly consists of bullnose particleboard treads, 2x12 dimensional lumber stringers, ¾-in. OSB risers, and ¾-in. scrap ripped to make padding strips for drywall.

Stair parts	Material option	Cost
Treads	⅝-in. x 12-ft. bullnose particleboard	$12.65
	Two 12x12s	$12.63
	1¼-in. x 11⅞-in. x 12-ft.-long LSL	$19.66
Risers	¾-in.-thick 4x8 OSB yields 14 risers	$11.81
	1x8, 12 ft. long, yields 4 risers	$7.64
Stringers	2x12, 18 ft. long	$16.24
	1¼-in. x 17½-in. LSL	$34.99
Padding for rock	1x4, 16 ft. long	$5.72
	¾-in. OSB scrap ripped to 3½ in.	$0.00

CUTTING WASTE ON STAIRWAYS

Locating stairways carefully to coincide with a 4-ft. building module allows the use of less framing lumber. Where the stairway header remains within 3 ft. of the end of the joist span and the opening does not exceed 4 ft., the opening can be framed with single joists and headers without hangers.

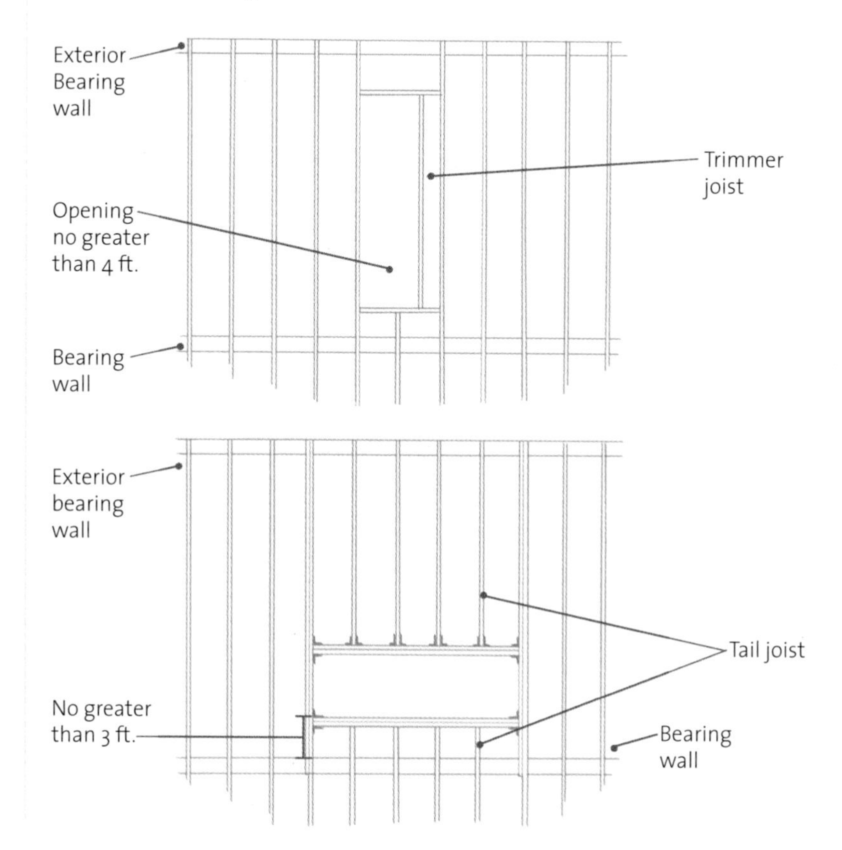

bay. This arrangement preserves structural joists and allows the use of stud scraps for headers instead of wasting more expensive material.

Stretching stair stringers

Although it's traditional to build stairs with three carriage boards, you don't always need them. Even when you use a 9½-in. board, if you design your stairs with a tread width of 36 in. or less, you can use just two stringers by limiting your maximum run and rise to 4 ft. 6 in. and 4 ft. respectively. To extend these dimensions further, you can nail a continuous 2x4 board along the bottom edge of each stringer, thereby stretching their capacity to a run of 5 ft. and a 4-ft. 9-in. rise.

> Stairwells can consume many construction dollars if you're not careful.

Since most stairways span a full story and not just 5 ft., you can still avoid the third stringer in longer spans by dividing the run into two segments with an intermediate support. Tread widths of 42 in. and wider usually will require a third stringer.

Instead of attaching the high end of the stringers to the header with framing anchors, build stairs on the ground and nail a sheet of ½-in. OSB to the end of the stingers. Nail this OSB ledger to the header. There's no need for heavy-duty connectors at the head of the stairs because stringers exert their load at the bottom; they don't "hang" from the high-end header. Just think of an extension ladder—do you want someone to steady it from the bottom or the top? At the low end of the stringers, a simple 2x4 ledger notched into the carriage boards provides a solid anchor to arrest the weight of the steps.

Although it costs a little more, you should consider using 1¼-in. TimberStrand LSL rim boards for full-flight stair carriages. This engineered product provides clear structural benefits, including no shrinkage and better nail-holding power, both of which can help eliminate stair squeaks and loose treads.

USING TWO STAIR STRINGERS

If stairs are kept to 3 ft. wide, it's possible to use a two-stringer design. For longer stairways, instead of adding a third stringer, split the run in two by framing a midspan pony wall.

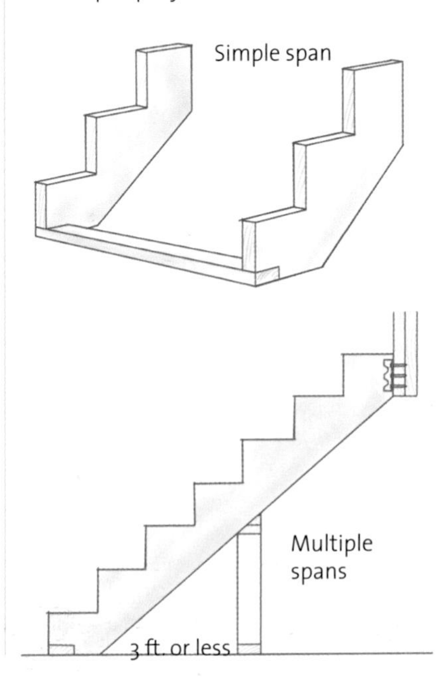

Two-Stringer Staircases

If the inside dimension between stringers is 36 in. or less, and the outside dimension of the step is 40 in. or less, a two-stringer stair assembly can be used. This chart shows the maximum rise and run for residential two-stringer staircases.

Material		Without reinforcement	With 2x4 reinforcement
2x10	Total run	6 ft. 0 in.	6 ft. 9 in.
	Total rise	5 ft. 8 in.	6 ft. 5 in.
2x12	Total run	9 ft. 0 in.	10 ft. 6 in.
	Total rise	8 ft. 7 in.	9 ft. 7 in.
2x14	Total run	12 ft. 0 in.	12 ft. 0 in.
	Total rise	11 ft. 3 in.	11 ft. 3 in.

OSB Makes Economical Risers

Instead of leaving a 1½-in. gap between steps and an adjacent wall for a finished skirt, eliminate this expensive piece of trim and leave a ¾-in. gap between the steps and wall. The gap leaves plenty of wiggle room to insert drywall, and instead of using a stud to create the space, you can rip a 4-in.-wide strip out of leftover floor sheathing.

You can save even more money by using leftover floor sheathing to make risers instead of buying 1x8 boards. An average stairway has 14 risers. If steps are 3 ft. wide, these stairs will require six 8-ft.-long boards and leave you with a waste factor of 2 ft. along with an extra riser. From a single 4-ft. by 8-ft. sheet of OSB, you can cut exactly 16 risers without any waste at all. You can use the two extra risers on the next set of stairs, but even if you don't, the difference in cost between a sheet of ⅝-in. OSB and six pine boards is nearly 50%.

Floor sheathing won't work on steps if stringers are more than 24 in. apart, but you can buy 1¼-in. by 12-in. high-density particleboard bullnose stepping, which provides a solid, dressed surface comparable in stiffness to dimensional lumber. Particleboard steps cost less than 2x12 treads and provide a more reliable step that won't crack or warp. What's more, particleboard stepping comes in 7-ft., 9-ft. and 12-ft. lengths, so you can buy exactly what you need with little or no waste.

Although particleboard treads are more economical than 2x12 lumber, they still cost more than 2x10s, especially the ones lying around the job site. Once covered with carpet, there's no advantage to particleboard. I always prefer to use scrap lumber for things like treads and risers rather than throw good wood into the garbage bin. But whenever I don't have any extra boards on the job—or if I want the extra depth and nosing a 12-in. tread provides—I always choose particleboard steps over dimension lumber.

ATTACHING STAIR STRINGERS

Instead of using framing anchors to attach the tops of stair stringers to a header, try using a gusset made from OSB. At the base of the stair, nail a 2x4 block between stringers and attach it to the subflooring.

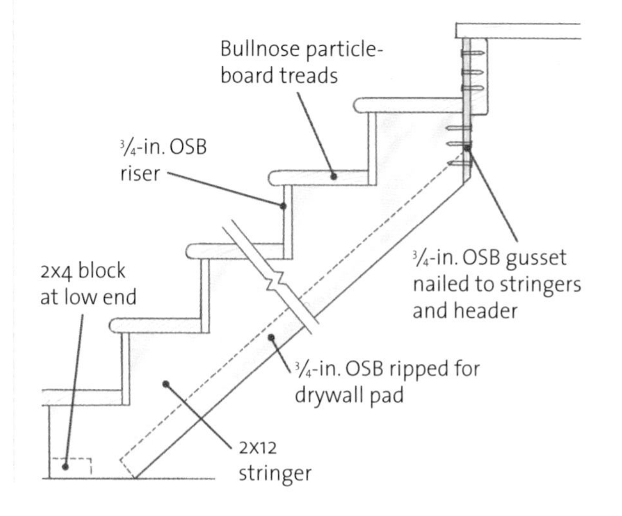

Weyerhaeuser's LSL has become a popular material for stair stringers, but 2x12s cost about half the price. In full-length stairways, where shrinkage in dimensional lumber stringers can cause squeaks, the extra cost may be worth it.

TRADE SECRETS

When using a steel beam, it's not necessary to lay a ledger board over the beam before toe nailing joists. Just drive a 10d sinker into each side of the joist, and then clinch it over the steel girder flange.

Girders, Beams, and Posts

Girders, beams, and posts provide midspan support for floors where there is no bearing wall. For maximum benefit in reducing joist spans, it's usually best to locate girders at the centerline of the structure. To save material, however, it sometimes pays to place the girder up to 12 in. off center. For example, if the building is 26 ft. wide, rather than place the girder at the centerline and use two 14-ft. joists to cover the 13-ft. span on either side of the girder, offset the girder 12 in. and use 12-ft. joists on one side, and use 14-ft. joists on the other.

For maximum benefit in reducing joist spans, it's usually best to locate girders at the centerline of the structure.

The type of girder or beam you choose depends on cost, headroom, and loading conditions. Reducing the cost of either requires minimizing the beam size and maximizing the column spacing. Steel and engineered-wood beams are usually more expensive than lumber, but permit longer spans and fewer column supports. When you factor in labor, the decision to use an alternative to dimension lumber sometimes results in a better bottom line.

For shorter spans, site-built wood beams made from standard dimensional lumber with a sheet of ½-in. plywood sandwiched between are generally the most cost-effective approach. But as spans get wider than 12 ft. in a single-story home, or 10 ft. in a two-story, the option of using steel or an engineered-wood product becomes more profitable, especially if the beam will be covered with drywall. Steel and engineered-wood products are stable and won't shrink. When lumber shrinks, drywall cracks and nails pop.

Handling headroom concerns

Even though it may be structurally adequate and less costly, a standard built-up beam may be too deep to provide enough headroom. For example, if the ceiling is 7 ft. 6 in. high, a built-up beam made from 12-in.-deep boards may drop the ceiling too much. Most codes require a minimum of 80 in. of headroom. One solution is to lift the beam into the floor framing and use metal hangers to attach the joists. But it's a bad move, so don't even think about it.

Avoid flush girders if possible

Unless it's absolutely necessary to have a flush ceiling, it's almost always less expensive to drop the beam and place joists on top of it, even when this requires purchasing a steel or engineered-wood girder to achieve adequate headroom. Cutting joists precisely to length, purchasing hardware, and installing hangers adds substantial costs to any beam installation.

Girder Spans

Sample of allowable spans for built-up wood-center girders:

Width of structure	Girder size	One story	Two stories
28 ft.	Three 2x8s	6 ft. 3 in.	4 ft. 5 in.
	Three 2x10s	7 ft. 7 in.	5 ft. 4 in.
	Three 2x12s	8 ft. 10 in.	6 ft. 3 in.

Steel Girder Spans

Sample of allowable spans for steel-center girders:

Width of structure	Beam size	One story	Two stories
28 ft.	W 6x9	11 ft. 4 in.	8 ft. 10 in.
	W 10x12	16 ft. 10 in.	12 ft. 4 in.
	W 14x22	26 ft. 0 in.	20 ft. 0 in.

Steel is heavy, but a 9-in.-deep steel girder can replace four 2x12s. If a crane will be on the job site anyway, steel is worth considering for applications where girder height is critical.

Steel girders stronger but heavy

Steel girders cost more than wood and sometimes require a forklift or crane to set in place, but if the equipment is available on site, consider the advantages of a typical 9-in.-deep steel girder. It can replace four 2x12 boards put together with nails and glue.

When using a steel beam, it's not always necessary to use a 2x6 ledger board for toenailing joists. Instead, place your joists directly over the beam and drive a 10d sinker into each board, then clinch it onto the steel girder flange. The main reason to avoid steel is to avoid the hassles of an additional supplier. This is why many builders prefer to use engineered-wood girders supplied by the lumberyard.

The advantages of engineered wood

Engineered-wood products—such as glue-laminated beams (glulams), laminated-veneer lumber (LVL), and parallel-strand lumber (PSL)—appear poised to replace traditional dimensional lumber for beams and headers. Although environmentalists have reservations about the toxicity of glues used to produce these engineered beams, the diminishing supply of old-growth timbers has forced lumber interests to develop ways of using rapid-growth species as a replacement.

Manufactured lumber is essentially reassembled lumber with its natural defects—like knots and holes—removed to produce a uniform member with precisely engineered characteristics. Engineered girders have uniform structural properties and a lower moisture content that makes them less susceptible to shrinking and warping. They can also span greater lengths than standard dimension lumber, partly because they have reliable performance standards that reduce the margin of safety required in design.

From a cost perspective, these features allow you to specify your girders and beams more precisely and to increase the spacing

Engineered-wood products, a broad category of materials that includes wood I-joists and other structural components, is overtaking sawn dimensional lumber in residential construction for some applications.

between column supports. When you consider the cost difference between a built-up beam and an engineered girder, don't forget to factor the labor required to construct a built-up beam, along with the glue, nails, and, sometimes, even nuts and carriage bolts.

Of all the manufactured-beam options, LVLs are generally the best type of engineered lumber for girders. LVLs cost less than glulams and provide much greater load capacity. Although PSLs are stronger, they are only marginally stronger for typical residential spans and cost considerably more than LVLs. For short, header-length spans, consider LSL, which costs less than other engineered lumber products and works well for less demanding applications.

Of all the manufactured-beam options, LVLs are generally the best type of engineered lumber for girders.

Making plywood beams

Plywood I-beams and box beams represent other bright design ideas that came out of NAHB research into OVE framing. You can manufacture plywood beams on site using dimension-lumber plywood, nails, and glue. The NAHB (*Beam Series*, NAHB Research Center, 1981) and the American Plywood Association (*Plywood Design Specifications*, 1998b) both publish span tables and engineering guides for these convenient and inexpensive site-built beams. The IRC also includes a simplified "cookbook" of allowable spans for 9-in.- and 15-in.-deep wood structural panel box headers.

Known in the trade simply as box beams, the most popular variety of site-built plywood girder consists of two continuous dimension-lumber flanges (typically 2x4s or 2x6s) sandwiched between two plywood webs. A box beam can be built with only one plywood cover, but it has considerably less strength. The IRC prescribes maximum allowable spans for both.

Short box-beam spans don't require intermittent support, but in widths of 4 ft. and greater—as most girders are—cripple studs will have to be added on 16-in. or 24-in. centers, depending on the wall-framing pattern. At both end-bearing points and at any intermittent point loads, you'll have to install double-cripple-stud stiffeners. If you're using a box beam in an exterior wall, you'll have to add insulation between the cavities.

These beams are most useful in girder design because they are the least expensive code-prescribed option for bearing spans from 4 ft. to 8 ft. The extra width created by the plywood webbing makes them less practical as in-line wall headers.

Another option is to build a wood and plywood I-beam with a single panel sandwiched between two dimensional-lumber flanges with cripple-stud stiffeners added just as in the box-beam construction. Although you can find span tables for I-beam construction, the IRC does not provide prescriptive guidelines.

Box-Beam Spans

Sample of allowable spans for wood structural panel box headers (box beams):

Width of structure	Header construction	Header depth	Allowable span
28 ft.	Wood structural panel 1 side	9 in.	3 ft.
	Wood structural panel 2 side	9 in.	5 ft.
	Wood structural panel 1 side	15 in.	4 ft.
	Wood structural panel 2 side	15 in.	7 ft.

Built-up 2x4 columns are a good choice

Columns used to support wood and steel girders range in cost from built-up 2x4s to lally columns. By far the least expensive are sandwiched-wood studs because none of the alternatives provide significant labor-saving advantages. Columns just aren't that difficult to install.

Although steel columns have a theoretical advantage in length and maximum spacing, few residential structures can exploit these benefits. Most homes don't have spans of 25 ft. or more between supports. The advantages of steel aren't as meaningful in residential construction columns as with beams.

Wooden columns can be single members, such as a 4x4, or site-built combinations of three studs nailed together. As a stand-alone support, a built-up column does not provide the same structural capacity as a sawn column of the same nominal size. But if buried in sheathing or drywall, a built-up column can actually perform better. Built-up columns are also less costly because there are generally a few extra studs lying around. This is especially true with larger-size supports, such as a 4x6 column. Ask your engineer to specify nailed-together, built-up columns whenever possible. If your engineer requires bolts, which entail a great deal of labor to install, then you may want to forgo this idea and use a single sawn member.

Just as with beams, lumber manufacturers have developed engineered columns, such as Weyerhaeuser's TimberStrand members. A 3½-in. by 3½-in. TimberStrand column has about one-and-a-half times the strength of a sawn column of the same dimension. Since engineered lumber costs more than standard sawn lumber, this is not an apparent cost advantage. But if the load exceeds the capacity of a sawn 4x4 column, pushing you into a 4x6 member, the engineered product actually costs less. The only way to know is to read the span tables or ask an engineer to calculate loads and provide you with design alternatives that you can price at the lumberyard. Some lumberyards have in-house engineering services that can provide you the information free.

Installing Subfloors

In the past, conventional floor sheathing was a double-layer assembly including subfloor and underlayment. Most builders now reject this approach in favor of a single layer of underlayment-grade sheathing. Although some builders continue to install a separate sheet of underlayment under resilient flooring, this is not necessary with certain loose-lay sheet goods (check with your flooring supplier). Eliminating the cost of an extra layer of underlayment can save roughly $300.

> Eliminating the cost of an extra layer of underlayment can save roughly $300.

One of the reasons builders still resort to additional underlayment is soiling and damage to the subfloor during construction, especially from drywall mud. Cleaning up thoroughly enough to lay resilient flooring can be more troublesome and time-consuming than covering the area with an extra sheet of underlayment. To avoid this dilemma, protect your single-layer sheathing with construction paper before putting up drywall. This keeps the sheathing clean and dry until you're ready to install your flooring.

Install subfloors correctly to silence squeaks

According to a survey conducted by the NAHB, one of the most common homebuyer warranty complaints involves "spongy" or squeaky floors. Builders usually try to reduce these callbacks by using larger dimension joists, cross bracing, and proprietary framing systems that stiffen the

TRADE SECRETS

To avoid floor squeaks, try screwing your subfloor to the joists through a resilient membrane, such as the Integrity Gasket™. It reduces impact-sound transmission through floors and walls, and reduces floor vibration and squeaks.

floor framing and improve stability. But properly installed sheathing is probably the best and most cost-effective solution to squeaky floors. When glued and fastened to joists with deformed shank nails or screws, sheathing combines with your floor frame to create a T-beam that not only mitigates squeaks, but provides a more solid surface. Upgrading the sheathing usually results in a more silent and sturdy floor than upgrading floor joists.

> From a value-engineering standpoint, it's wise to calculate the cost difference between narrow joist spacing with conventional ⅝-in. sheathing and wider joist spacing with thicker sheathing.

Choosing the right floor sheathing

From a value-engineering standpoint, it is wise to calculate the cost difference between narrow joist spacing with conventional ⅝-in. sheathing and wider joist spacing with thicker sheathing. Tongue-and-groove OSB can span up to 48 in. without intermittent support.

If joists are spaced at 24 in., it's advisable to install ¾-in.-thick structural floor panels. Use plenty of construction adhesive and 10d deformed shank nails at 6-in. centers on the edges and 10-in. centers in the field to create a uniform bond between the sheathing and joists.

Because dimension-lumber joists and prefabricated wood I-joists (see next section on engineered floor

Wood I-joists weigh about half as much as sawn dimensional lumber and can be ordered in lengths up to 80 ft.

Joist Spacing and Subfloor Combinations

Maximum joist spacing table for various thickness of subfloor underlayment:

Span rating for tongue-and-groove underlayment	Nominal thickness	Maximum joist spacing with edge support or tongue-and-groove
16 in.	7/16 in.	16 in.
20 in.	19/32 in.	19.2 in.
24 in.	23/32 in.	24 in.
32 in.	7/8 in.	32 in.
48 in.	13/32 in.	48 in.

systems) are inexpensive, the optimal floor framing assembly consists of 20-in. to 24-in. joist spacing with ¾-in. glued-and-screwed single-floor OSB panels. But if floor trusses are used, which can cost considerably more per piece than conventional joists, check into thicker OSB single-floor panels that can span up to 48 in. and save you as much as 40% on your floor joists.

Engineered Floor Systems

Following the lead of roof-truss manufacturers, which have taken over the market from conventional stick-framed builders, the forest-products industry has developed a host of proprietary pre-engineered floor-framing systems. The most popular consist of wood I-joists, composite-wood joists, and parallel-chord floor trusses.

The principal advantages of a proprietary engineered floor system include improved structural stability, longer joist spans, ease of assembly, and, in the case of floor trusses, greater accessibility for wiring, pipe, and mechanical runs. Their disadvantages are higher cost, lack of standardization, and, in the case of steel, more difficult construction.

Sometimes higher material prices can be offset with lower labor costs. Manufactured joists, such as wood I-beams, are light enough for one man to handle and come in lengths that can span most residential widths without laps or splicing. Unfortunately, framing subcontractors don't usually provide a price break for engineered floor framing, so the homebuilder does not realize any price advantage. However, the larger spans permitted by pre-engineered systems may provide a marketing advantage since they allow greater design flexibility.

Sometimes higher material prices can be offset with lower labor costs.

Wood I-joists are now common

Wood I-joists are the most widely used engineered floor framing system today, representing approximately 30% of the total number of residential floors framed in 2002. Produced by numerous manufacturers, wood I-beams consist of two machine-stressed 2x4s, or LVL flanges, grooved to receive an OSB or plywood web glued in place. The most widely known is Weyerhaeuser's TJI®, part of the SilentFloor® system.

Advocates claim that lower labor costs offset the higher price of these engineered joists. In a typical house, this difference in price can add up to several hundred dollars. In the small, narrow

Engineered floor systems, which may include wood I-joists, have several advantages over conventionally framed floors, including better structural stability, longer spans, and ease of assembly.

Use of Steel Joists Limited

Steel framing systems are common in commercial applications, but less than 1% of all single-family homes use any steel framing at all. In an effort to market their products to homebuilders, steel manufacturers have developed products that dovetail with traditional wood framing. One of the most successful applications of steel components, developed by Dietrich Industries®, is the TradeReady® steel floor system.

These joists have 6¼-in. by 9-in. oval knockouts to facilitate electrical and mechanical runs. The large openings only come in the deepest and most costly engineered-wood products. Because of the inherent strength of steel, the Dietrich product is small enough in standard residential applications to compete in price with wood I-joists. The greater span capacities and deflection properties of steel also allow for wider spacing, such as 32 in., without requiring a tradeoff in excess depth, as with floor trusses.

Although the TradeReady steel floor system was designed for framers to install, most carpenters who have tried them report higher labor costs and a marked learning curve in adjusting to steel, especially when fastening floor sheathing. In order to comply with code, electricians must install plywood gussets over the knockouts and then drill holes as usual to pull cable. Just like wood trusses, these joists represent a specialty product for certain upgrade markets, but they are not yet a viable cost-savings solution for the affordable-home builder.

TradeReady Spans

Sample spans for TradeReady® steel joists.

Joist member and nominal depth	16-in. spacing	19.2-in. spacing	24-in. spacing
TDJ18 (7.25 in.)	13 ft. 4 in.	12 ft. 7 in.	11 ft. 8 in.
TDJ18 (8 in.)	14 ft. 5 in.	13 ft. 7 in.	12 ft. 6 in.
TDJ18 (9.25 in.)	16 ft. 3 in.	15 ft. 2 in.	13 ft. 7 in.
TDJ16 (11.25 in)	20 ft. 6 in.	19 ft. 4 in.	17 ft. 11 in.

homes that I build, I haven't found sufficient labor savings to justify the extra expense. In large homes, the added span capacity of engineered joists can eliminate the need for midspan support.

Some brands come with prescored knockout holes to facilitate electrical and plumbing installation. These holes are not big enough for anything but the smallest distribution ducts, so you still have to drop ceilings or build chases for the air handling system.

Open-web floor trusses are a pricey alternative

The use of prefabricated open-web floor trusses has become widespread in multifamily construction because of their long span capacity and flexibility for mechanical access. Their use in single-family-home building is growing, but while they offer certain quality and construction advantages, they add considerable expense.

Capable of spanning distances that exceed most residential applications, open-web trusses are light and easy to handle. Open-web trusses can easily span foundation walls when freezing temperatures make it impossible to install a floor slab, allowing construction to advance even when it's too cold for concrete. Sometimes, it's just more cost-effective to keep working than to take a winter vacation.

Wood trusses provide plenty of room to run plumbing, electrical, and other mechanical lines.

Open-web floor trusses can span long distances and are widely used in multifamily housing. But they also cost more than many other alternatives.

Since a truss plant has to design and assemble every floor system individually, floor trusses arrive at the job site cut to length and ready to install. The large web openings between flanges make mechanical installations especially easy. Sometimes the openings can provide enough space for an air-conditioning plenum. Open-web trusses are the most construction-friendly floor-framing system available. Unfortunately, it can take considerable lead time to order trusses, and, in most residential applications, you can't make up for their high cost with labor savings.

> Open-web trusses are the most construction-friendly floor-framing system available.

A new option from Truswal

The Truswal Systems Corp.® has developed a hybrid of wood I-joists and open-web trussing that provides a less costly alternative. It's SpaceJoist TE™, which does not require custom manufacturing. SpaceJoist TE sells like wood I-joist systems, with a job-specific design assembled from mass-produced components, which means there's no lead time required. Just order and build.

Although the system costs about a third more than wood I-joists, I can make up for this difference by negotiating a small price reduction with my framing, electrical, plumbing, and HVAC subs. These trusses are light, easy to install, and make the job of running wires and mechanical systems very easy. Framers don't have to come back after the subs have finished their work to install chases and repair overcut joists.

One reason these joists cost less than traditional floor trusses is that you can work with standard-length joists and trim them to fit your span. At either end of a SpaceJoist TE is a solid web of OSB—just like a wood I-joist—that provides 12 in. of trimmable area. Between the

SpaceJoist TE combines the best of wood I-joist design with the advantages of open-web trusses. These lightweight joists come with 12-in.-long ends that can be trimmed, an open-web design, a wide nailing surface, and a built-in plenum chase.

I-joist ends is a steel open-web design. These joists come with a built-in plenum chase measuring 6¼ in. by 24 in. in a nominal 10-in.-deep joist. Larger joists have even larger plenum openings. Standard open-web trusses can't accommodate an HVAC plenum in a 10-in.-deep joist. The SpaceJoist TE can cantilever up to 2 ft., just like a dimensional-lumber joist.

TrimJoist™ is a contractor-friendly all-wood truss system that allows you to trim the product on site. Many other products in this category should be available soon.

Framing the Roof

Throughout most of the United States and Canada, prefabricated roof trusses have displaced conventional roof framing. Trusses can be installed quickly, require less carpentry expertise, and can accommodate almost any roof configuration imaginable. But in some situations, roof trusses cost more than conventional framing, and many builders find that optimizing roof construction sometimes requires a blend of prefabricated components and old-fashioned rafters.

Throughout most of the United States and Canada, prefabricated roof trusses have displaced conventional roof framing.

Conventional framing vs. trusses

As with floor trusses, the advantages of prefabricated roof components include engineering, flexible span options, and ease of installation. Each time you erect a single roof truss, you've completed three operations in one: rafters, joists, and bracing. It takes a three- to five-person crew to build a conventional roof, while one carpenter and a helper can install the same roof with trusses. Since the manufacturer can engineer most trusses to span between outside walls, you can sometimes eliminate girders and interior bearing walls completely, saving money on wall framing. Trusses accommodate advanced framing techniques because they usually install at 24-in. spacing.

On simple roofs, conventional framing can still provide a cost benefit, especially if the local truss manufacturer requires long lead times. To optimize conventional framing, use a 24-in. module, choose rafter and joist sizes based on actual loads, and use the least expensive species of wood available that codes and span tables permit.

One of the advantages of conventional framing is the ability to construct attic rooms or storage areas and dormers. Attics and dormers are notoriously expensive with trusses. This is where a combination of manufactured trusses and conventional roof framing is a more cost-effective solution than either system alone.

Comparing the Cost of Joists

The cost comparison of floor joist types and optional layouts on one sample floor 28 ft. by 40 ft. with a midspan support is illustrated below. Although 2x6s at 12 in. on center appear to be the most cost-effective, 12-in. joist spacing works against the mechanical trades and could trigger a price increase. The open-webbed SpaceJoist TE might actually provide an incentive for negotiating lower costs with your mechanical trades. Overall, the best economy of materials and workability comes from the often-recommended 2x10 joist spacing at 24 in. on center.

Material	Spacing	Cost	Sheathing	Cost	Total cost
2x8x14	16 in o.c.	$518.75	⅝ T&G OSB	$426.82	$945.57
2x10x14	24 in. o.c.	$534.33	⅝ T&G OSB	$426.82	$916.15
2X6x14	12 in. o.c.	$462.42	⅝ T&G OSB	$426.82	$889.24
2x12x14	32 in o.c.	$512.96	¾ T&G OSB	$476.67	$989.63
9½ in. NASC	19.2 in. o.c.	$690.46	⅝ T&G OSB	$426.67	$1,117.13
9½ in. TJI	19.2 in. o.c.	$1,202.00	⅝ T&G OSB	$426.82	$1,628.82
9½ in. TJI	24 in. o.c.	$1,006.00	⅝ T&G OSB	$426.82	$1,432.82
SpaceJoist TE	24 in. o.c.	$1,772.40	⅝ T&G OSB	$426.82	$2,199.22

It can take a large crew several days to frame a conventional roof. Using prefabricated roof trusses, two people can install the same roof in a morning and start sheathing in the afternoon.

Many builders have taken to assembling wood trusses on the ground and later craning them onto the roof in large, pre-assembled sections. If you have the equipment available, this makes the assembly of roof trusses much easier and quicker.

To make attic construction easy and affordable, the staff of the NAHB Research Center used a combination of trusses and simple conventional framing to build a family room and ample storage space into this demonstration project in Bowie, Md.

An alternative to attic trusses

In areas where homes typically don't have basements, attics have become the preferred method of providing flex space. Unfortunately, attic storage and bonus rooms are expensive to build. Truss manufacturers can provide ready-made roof framing with an attic area built in, but you save a lot of money by building a three-part system with site-framed cripple walls (or truss girders), scissor trusses, and conventional rafters.

Prefabricated gable ends are faster

When you buy a truss package, you generally purchase a pair of prefabricated ends, too. These are similar to trusses, but instead of a structural web, they have vertical members at 24 in. on center for nailing on siding and sometimes for interior drywall. This saves considerable time and expense over building rake walls or cutting and fitting cripple studs under a rake rafter.

Prefabricated gable ends don't have the structural capacity of a web truss and cannot span unsupported areas. For the most part, gable ends don't need to be structural or even

CHEAPER THAN ATTIC TRUSSES

A three-part attic truss framing system used by the NAHB Research Center on a demonstration home in 2002 is less expensive than buying a specialized attic truss. Although it requires assembly, it was actually easier to install. No crane was required to hoist heavy trusses.

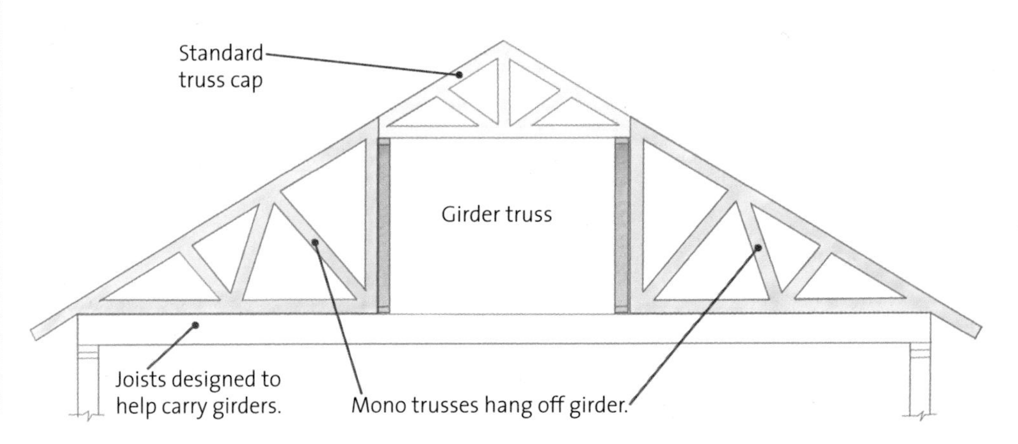

self-supporting. But if the gable-end wall has a wide opening—such as a garage door that requires a large structural header—you can take advantage of the bearing capacity of field trusses to take the load off the opening by using a standard truss at the gable end and furring out the web with flat studs to receive cladding. Or, the manufacturer can provide a girder truss that combines a structural web with vertical nailing flanges. This will cost a little more than one extra field truss and a few scraps of lumber.

There is no practical argument for an overhang at gable ends.

Simplify overhangs and cornice details

In most regions of the United States, overhangs are desirable but not required. A traditional boxed cornice can take longer to build than the rest of the roof. Although there is an argument that a soffit protects walls and windows from sun and rain, there is no practical argument for an overhang at gable ends. If you look at houses in older neighborhoods, you'll find many examples of roofs with small soffits and no gable-end overhangs at all. Of all the ways to cut costs on your roof, simplifying your cornice details is one of the most profitable.

I have been building homes without overhangs on side and rear elevations for about 12 years and have never had a leak or moisture problem. One of the reasons for this success is that I clad my fascia board with aluminum siding, creating a Z-bar® flashing that tucks under my drip edge on the roof and over the house wrap on the wall. A gutter finishes the assembly along my rafter-end elevation.

In the South and along the West Coast, builders often use an exposed-rafter-tail soffit that's easy to install, sheds water, and provides ample shade. With today's exterior-grade sheathing materials, there's no reason not to use this design anywhere in the United States.

Although it's not necessary, you may want to install a box overhang at the rake edge for aesthetic reasons. I install them at the front elevation of my homes to conform to

Roof overhangs are common, but in most parts of the country they are not actually necessary and only add to the complexity and cost of framing a roof.

This traditional saltbox house has 8-in. soffits and no gable end overhang.

A zero-overhang aluminum-clad fascia under construction. A gutter will finish the assembly.

DETAILING THE ZERO-EAVE OVERHANG

Aluminum installed over the fascia and lapped over the housewrap on the wall helps keep the assembly weathertight even without a protective overhang.

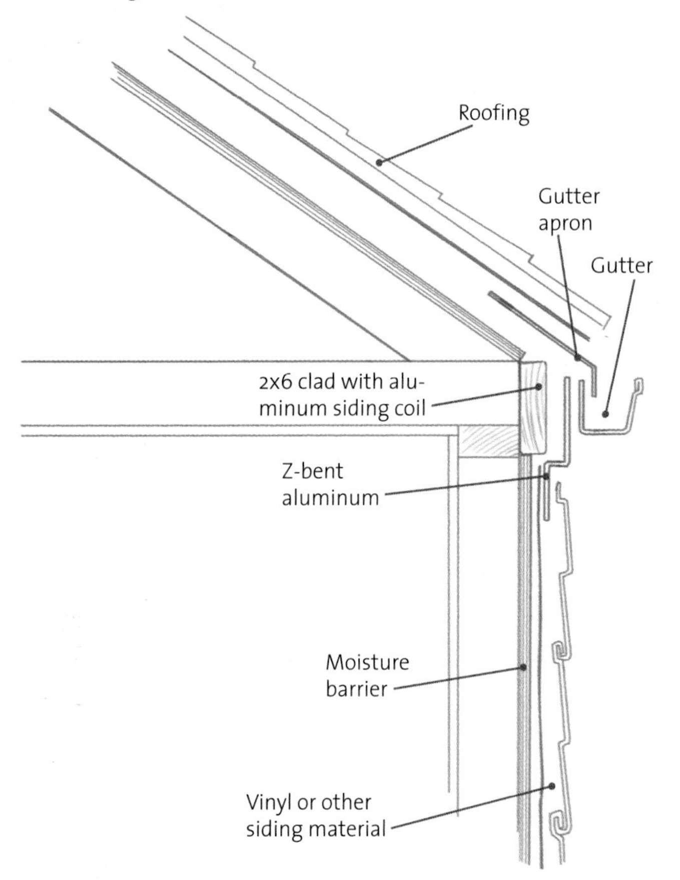

Fire-Resistant Eaves

Fire-resistant roof eaves are one of the more difficult assemblies to build. But by using stucco on the house, you can build up a "layer cake" of 2x6 and 2x4 material to create a stepped profile at rakes and soffits and then wrap this with three-coat stucco for an easy, appealing one-hour fire-rated eave. You can accomplish something similar by building a 6-in. to 8-in. eave skinned with fiber-cement siding.

Minimum Roof Overhang Widths

Climate index	Eave overhang	Rake overhang
Less than 20	N/A	N/A
21 to 40	12 in.	12 in.
41 to 70	18 in.	18 in.
More than 70	24 in.	12 in.

Adapted from *Prevention and Control of Decay in Homes* by Arthur F. Verrall and Terry L. Amburgey, prepared for the USDA and HUD (Washington, D.C., 1978).

local custom. It's easiest to build a box rake on the ground and then nail it in place as an appliqué. If your overhang does not exceed 12 in., toenailing along the rake and overlapping the roof sheathing is enough for a sturdy installation that you can walk on.

Many builders like to use ridge vents, which rely on soffit vents for air circulation. A modified soffit with rollout ridge-vent membrane can provide the needed inlet, but you have to be careful not to block this system with insulation.

Don't overlook roof ventilation

It's important to provide adequate ventilation to minimize the impact of moisture accumulation in the attic. Most builders use OSB sheathing, which swells with moisture and does not regain its original dimension once dry. This is why it's important to provide a ⅛-in. gap between sheathing panels. If there's no room for the panels to expand, the edges swell and buckle. You can see this effect in asphalt and fiberglass roofs, which telegraph the lines created by buckling panel

> It's important to provide adequate ventilation to minimize the impact of moisture accumulation in the attic.

edges. In extreme cases, the sheathing can actually become wavy.

With many of the overhang designs described here, it's not possible to use a soffit and ridge vent combination. But properly sized gable-end vents with roof-mounted pan vents just below the ridge have been in use for decades and still provide the most cost-effective solution to proper attic ventilation.

Most of the moisture problems found in roof systems relate to improper construction more than inadequate venting. To make sure you don't have attic moisture problems when using conventional vent systems, follow these simple guidelines:

> Don't exhaust bathroom or appliance fans into the attic.
> Install a moisture barrier in all of your attic ceilings.
> Install roof sheathing correctly, with gaps and midspan H-clips.
> Make sure to size and balance your roof vents for proper convection.

Most of the moisture problems found in roof systems relate to improper construction more than inadequate venting.

If you're going to use a ridge and soffit vent system, try the rollout ridge vents that are fastened with a nail or staple gun. These roll on quickly and have such a low profile that they rarely blow off. From the street, they're virtually invisible.

One way of reducing moisture problems in the attic and reducing your ventilation requirements is to install a vapor barrier in the ceiling. Building codes specify the minimum unobstructed vent area of a roof should be no less than 1/150 of the total insulated area of the ceiling. But if you place a vapor barrier on the warm side of your ceiling joists, the vent requirement drops to 1/300 of the total insulated area of ceiling.

The gable-end and pan vent combination is an old but useful method of providing adequate air circulation in an attic.

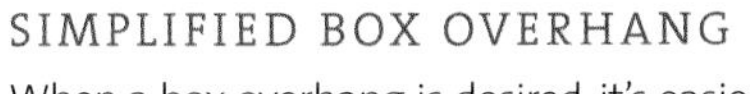

SIMPLIFIED BOX OVERHANG

When a box overhang is desired, it's easier to make it on the ground than it is to build it in place. When kept to a maximum of 12 in., this design is sturdy.

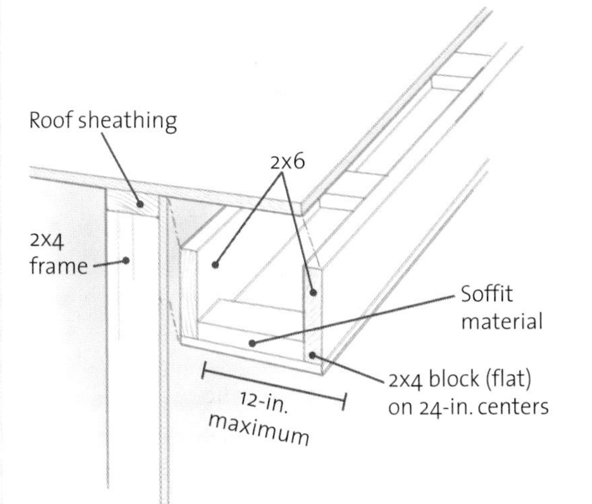

VENTING A ZERO-EAVE ROOF

Closed soffits are not compatible with ridge-vent systems. Here, a screen vent made from rollout ridge vent material is used to ventilate the zero-eave roof design.

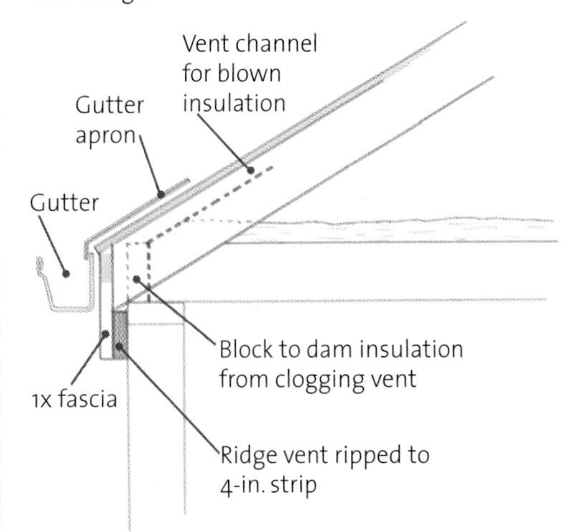

Split-flange foundation anchors, such as this USP lumber connector, secure mudsills and framing to concrete block walls, stem walls, and slab foundations.

Installing Framing Connectors

Few homebuilders pay much attention to the cost of framing hardware, but it is an area of surprising variety and potential savings. This is especially true for builders working in earthquake and hurricane zones where increasingly stringent requirements have driven the cost of framing hardware from a few hundred to several thousand dollars per home.

If you open up a full-line catalog of framing connectors, you'll find about 300 product types in several thousand sizes. Some of these can save money by making short work of complex cutting and fitting; others save money because they just cost less. In the latter category, you'll find many new designs for old standbys like shear-wall hold-downs. But all too often, builders keep using the more expensive version of the same product without realizing that the newer, less costly version has better engineering values and is easier to install.

Foundation anchors and hold-downs

Most builders already know to use split-flange foundation anchors instead of J-bolts. Although standard split-flange anchors cost about 30¢ more than H-in. J-bolts, they take less than half the time to put in and are more forgiving.

By replacing conventional anchor bolts, split-flange anchors eliminate time-consuming drilling and alignment problems associated with bolts. Testing by hardware manufacturers shows that nails provide a much bigger margin of safety. If there's a knot in the lumber where a nail is driven, there's little reduction in strength. There may be 15 other nails to distribute the load. But it compromises the entire assembly when this knot coincides with one of the bolts used to attach a typical hold-down.

Tension ties provide strong wood-to-concrete connections for anchoring shear walls in seismic and high-wind regions. United Steel Products® (USP), makers of the Kant-Sag® line of construction hardware, has developed a series of predeflected nail-in hold-downs that provide higher engineering values than traditional bolt-in brackets. When you include the cost of nuts and bolts, the two hold-downs cost about the same, but because of higher engineering values, you may use fewer of the predeflected hold-downs and then install them quickly with your nail gun.

Because engineers usually don't include cost into their intricate calculations, you have to help them select the right hardware. Review the structural plans and look at each hold-down that has been specified, then search for comparable products from competing manufacturers.

Save time with hangers and ties

Perhaps the most time-consuming and tedious task for a framer is nailing off joist hangers.

Nail-on hold-downs, such as this USP framing connector, attach more quickly and economically than either bolt-in or screw-in systems.

Hammering in special 1½-in. nails in a cramped joist bay can take hours. It's often left until last in the framing sequence—sometimes even forgotten. Slant-nailed hangers, however, can be installed quickly using your gun and common 16d nails. Because these nails have more shear value than traditional hanger nails, and they toenail through the joist into your beam, you don't have to shoot in quite as many.

When you're installing double I-joists at an angle, it's common to bevel-cut the joists to fit into a hanger. USP has developed a bracket that allows two-plied I-joists to hang staggered in the hanger, without requiring a bevel.

The "breakfast-nook" hanger carries four mono trusses to facilitate the installation of complex gazebo-style truss configurations.

In seismic and high-wind areas, builders have to brace their trusses laterally. Typically, this includes a diagonal gable-end brace from plate to ridge, fastened onto the plate with two framing clips. These gable-end braces can be very awkward to install. But USP has adopted an inexpensive and easy-fit HC520 clip to provide a one-step alternative.

Because of the bevel cuts, jack rafters have always challenged roof framers. One solution is a clever piece of hardware developed by USP to attach square-cut jacks at up to a 60° skew. At pennies over a dollar, these are preferred by many framers, and a less experienced carpenter can put them up quickly and without error.

In seismic construction, one of the more complicated and time-consuming tasks has been designing and building drag struts that connect a girder truss to an intersecting wall. For L-shaped homes, engineers have developed a simple connector to take the place of more rigorous framing. USP's drag strut connector transfers the load from a girder truss into a perpendicular bearing wall with a single, nail-on metal link.

Perhaps the most time-consuming and tedious task for a framer is nailing off joist hangers.

A product that some builders swear by while others find superfluous is the truss stabilizer. Mytech was the first to come out with it, but now several companies manufacture similar products. Although originally designed to provide fast and accurate spacing for roof trusses, it can be used for rafters and floor joists just as easily. The stabilizers come as multibay and single-bay spacers and install with one nail per framing member. They eliminate the need to mark modular layouts on bearing plates and provide temporary bracing until you install blocks and sheathing.

By using this USP 45° I-joist hanger, framers can stop bevel-cutting joists to fit.

Framing a breakfast-nook ceiling can represent a challenge unless you use this USP breakfast-nook hanger, which speeds up and simplifies the process.

The USP gable brace clip provides an easy connection for the truss brace.

Success Stories

Meet the Future

Imagine a framing subcontractor who provides lumber, builds his own trusses, and puts up a house in 72 hours. A builder's plans are fed into a computerized saw that selects its own lumber and cuts pieces so precisely that the total wood waste in a 2,000-sq.-ft. home fits into a single wastepaper basket.

This is Woodinville Lumber. The company, which employs 150 carpenters and is supported by a roof and floor truss plant, frames 1,400 to 1,700 homes a year in the greater Seattle area. Its four wall-manufacturing lines can cut, frame, and sheath an entire home in about two hours. Right next door, in a large, covered yard fitted with huge framing tables, carpenters cut and build floor panels to exact lengths and then load them on flatbeds for delivery. Across town, the Woodinville truss plant manufactures 25 roofs daily. All that's left to do on site is nail the pieces together.

> Well ahead of the curve, Woodinville may represent the future of home construction.

You can't buy a 2x4 from Woodinville Lumber. Although the company opened as a retail material yard in 1970, changes in construction economics prompted a move toward these specialized framing services. Two years after it opened, the framing department had grown to the point that Woodinville no longer needed retail lumber sales.

Well ahead of the curve, Woodinville may represent the future of home construction. As more lumberyards offer wall panels in addition to roof trusses, the promise of an integrated, engineered framing package becomes almost certain.

Fact Sheet

WHO: Woodinville Lumber

WHERE: Seattle, Wash.

WHAT THEY DO: Once a retail lumberyard, this innovative company produces framed wall and roof sections with state-of-the-art technology.

Plumbing: Less May Be Better

If you ask a real-estate agent what sells a home besides "location," he or she will tell you it's a big kitchen and a luxurious master bath. It's not surprising that many builders make sure their floor plans have outsized, wide-open kitchens and impressive whirlpool tubs. While this approach may work for the sales team, a frugal builder also appreciates that innovative plumbing design and material choices offer an excellent opportunity to reduce construction costs.

The Vanguard MANABLOC® manifold works like a circuit-breaker panel, centralizing all shutoff valves in one location.

You can save by using less expensive but better-quality distribution, waste, and venting systems along with a floor plan that optimizes materials and reduces the time needed to install them. The most expensive "fixture" in your house is the labor required to install pipes from one end to another. By economizing your plumber's labor and specifying easy-to-use

Keep Plumbing Fixtures Close Together

Plumbing codes vary throughout the country, allowing certain value-engineering alternatives (such as single-stack venting or plastic water piping) in one area but not another. Even in a city that remains staunchly conservative, certain universal methods for cost savings apply. Cluster plumbing represents the most basic approach—and it's relatively simple. It just means locating fixture groups, such as baths and kitchens, back-to-back or stacked over each other to save on drainage, venting, and water lines.

materials, you can save from $500 to more than $1,000.

To obtain these savings you'll have to persuade your plumber to bid on a time and material basis. Plumbers frequently use a per-fixture method of bidding, which can be inaccurate. I have found the best approach to reducing plumbing costs involves hiring your plumber as a cost-reduction consultant. Pay him up front to help you work on the redesign of your plans, and you won't have to struggle to obtain a price break later.

> The most expensive "fixture" in your house is the labor required to install pipes from one end to another.

It's a little tougher with the local plumbing inspector. The major code authorities approve all of the materials and approaches described in this chapter, but your municipality may not. Even so, become familiar with the alternatives and lobby local authorities to update the plumbing code. Cities throughout the nation update their plumbing codes every few years. By looking ahead, as soon as your local authorities catch up, you'll be ready to adopt these money-saving techniques. Meanwhile, you can still take advantage of the biggest cost-reduction strategy available: optimum layouts.

Smart Layout Cuts Costs

By minimizing the length of water and drain lines, you save money even if you're using traditional (and more expensive) cast-iron and copper plumbing. The first thing to find out

Plumbing systems offer often-overlooked opportunities for reducing construction costs through innovative design and material choices. Fewer supply and waste lines that run shorter distances are key, but it takes careful planning.

ONE TRENCH, NOT TWO

Placing water, sewer, and gas mains in a common utility trench saves money. The minimum distance between different types of lines required by building codes can be achieved with vertical separation.

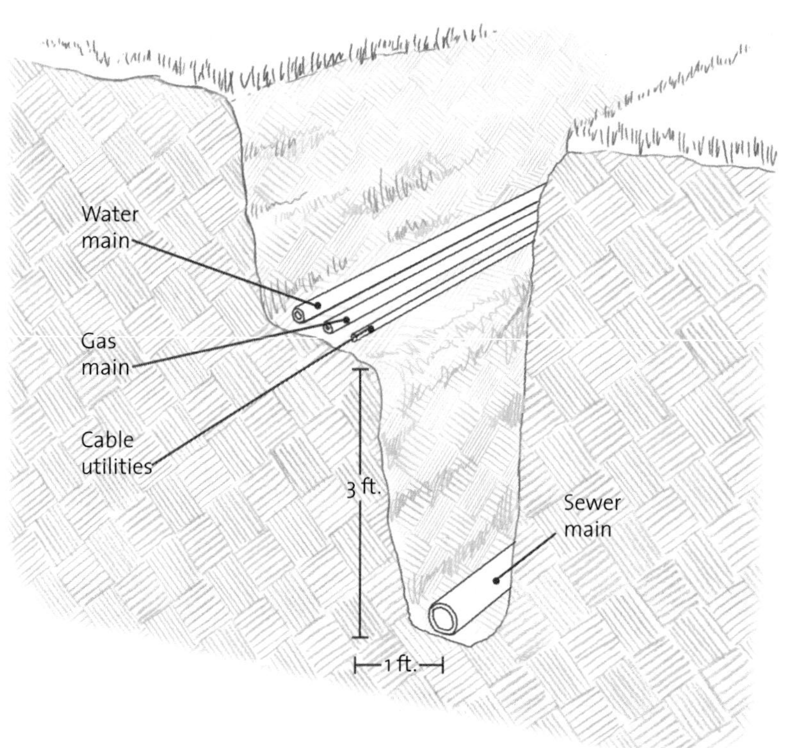

CLUSTER PLUMBING FIXTURES

Keeping plumbing fixtures in a small area minimizes supply, drain, and vent lines that must be installed. A floor plan that is not optimized for value requires more material and means higher labor costs for installation.

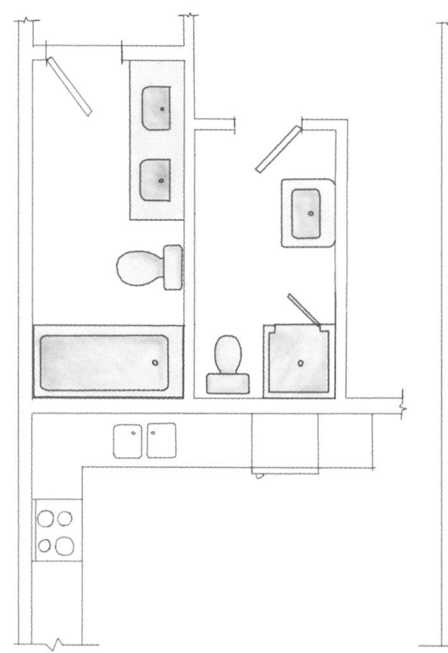

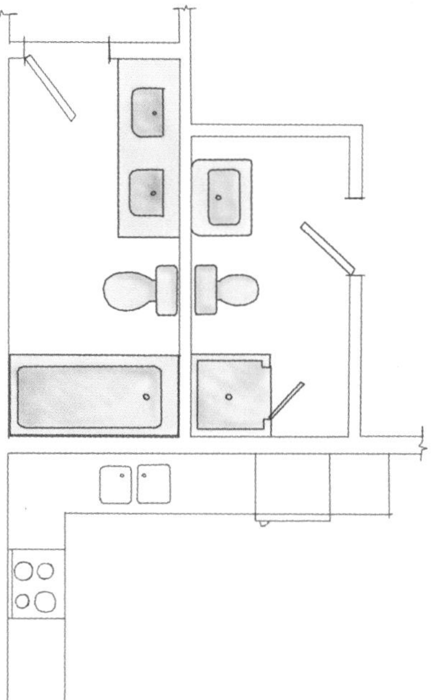

is the location of sewer and water service mains on the construction site. In my city, the service usually comes from the street side, so I try to locate bathrooms, the kitchen, and the laundry on the front of the house to reduce the length of service runs.

Most local codes require sewer and water service lines to be separated. If possible, use a common trench for sewer and water lines and separate them vertically to eliminate unnecessary cutting, excavating, and backfilling. Many building inspectors will allow this approach.

> Arranging the kitchen, bathrooms, and other wet areas back-to-back against a common wall concentrates piping in a compact area.

Keep plumbing concentrated

Arranging the kitchen, bathrooms, and other wet areas back-to-back against a common wall concentrates piping in a compact area. Use 2x6s to frame the wet wall, through which drain and vent lines can run with a little elbow room. Coordinating joists to avoid toilet waste lines and bathtub P-traps also saves time and trouble. This is especially true for engineered floor systems where tight restrictions on cutting and notching can make it practically impossible for your plumber to connect the drains.

Stack fixtures vertically

In houses of more than one story, stacking fixtures over each other eliminates horizontal runs and consolidates ventilation. Again, make sure to lay out joists and roof trusses so that the plumber has enough room to work directly above the wet wall. Whenever possible, provide a drop ceiling—that way, the plumber can avoid drilling and cutting joists to distribute waste and supply lines.

TRADE SECRETS

To avoid what I call market creep, which slowly increases the price of a house until it is no longer affordable, offer a very basic appliance package and a menu of upgrades. These upgrades can actually improve your bottom line if you add a reasonable markup. For example, a glass-top range may cost an additional $100, but you can usually sell this feature as a $150 upgrade.

Money-Saving "Green" Checklist for Plumbing

- Use ABS drain lines instead of PVC (see p. 108).
- Insulate hot-water lines.
- Place the water heater within the building envelope.
- Use CPVC or PEX water piping.

Copper pipe comes in flexible coils and rigid lengths with several wall thicknesses. Although some plumbers swear by copper, plastic is less expensive and usually performs better.

An Image Problem with Plastic

Plastic pipe suffers from an image problem related to polybutylene. The horror stories about leaky plastic pipe derive from problems with this product, which plumbers no longer use for residential drinking water applications. Modern plastic pipe based on CPVC or PEX doesn't break down with chlorine and no longer suffers from the manufacturing defects that plagued the old versions of plastic pipe. The modern plastic alternatives have many advantages, including ease of installation, a reasonable price, and a longer, more leak-free lifespan than copper.

Rethinking Supply Lines

Copper is still the most widely used pipe material for residential supply lines, but plastic is closing in. Most plumbers now use type M thin-walled copper for interior supply lines. The more costly type L copper is thicker and more resistant to chemical degradation, and most house service entrance lines are made with it. There's also a soft, flexible copper tube plumbers use primarily under house slabs because they can bend and loop it to fixture groups without leak-prone splices under the concrete floor. You can also use it aboveground to snake lines between floors, especially where joists are in the way. It's more expensive per foot, but it saves money in fittings and labor where you can't make a straight run.

Copper is still the most widely used pipe material for residential supply lines, but plastic is closing in.

Many plumbers swear by copper and, in the right environment, copper lasts for years. But copper has problems related to corrosion and leaching. In some areas with aggressive soil or water, problems with corrosion have become significant enough to force local building departments to ban type M and sometimes copper pipe altogether. For reasons of health, durability, and cost, builders are increasingly choosing plastic supply lines in their homes.

Although there's a perception that plastic is just cheaper, some builders offer certain plastic pipe systems as an upgrade. Plastic pipe designed for drinking water won't corrode, won't leach chemicals into the water, is less susceptible to freezing, has higher flow rates, and keeps hot water warmer than copper. It also costs about 20% less than type M copper and saves 40% to 50% in labor costs.

Chlorinated polyvinyl chloride

In use for nearly 40 years, CPVC (chlorinated polyvinyl chloride) is the most popular plastic piping for potable water. It costs about a third as much as copper, weighs less, cuts easily, and is assembled with inexpensive fittings and solvent welds. If you've ever installed a PVC (polyvinyl chloride) sprinkler system, you know already how to work with this material. The difference between CPVC and sprinkler pipe is that CPVC is rated for hot water. If your jurisdiction allows, you can actually save even more money by using CPVC for hot-water lines and regular PVC for cold.

The only caveat with this material is that it becomes brittle over time when exposed to sunlight. You can prevent damage from ultraviolet light by painting exposed sections of CPVC with exterior latex house paint.

Cross-linked polyethylene

The newest potable plastic water pipe is PEX (cross-linked polyethylene). As its name implies, the components in PEX are cross-linked under intense heat and pressure during manufacture. This process makes PEX resistant to considerable heat and deformation. Its outstanding feature is flexibility, which eliminates many of the joints necessary with a rigid system, such as copper or CPVC.

PEX dominates the market for under-slab heating and snow melt precisely because of its flexibility and resistance to freeze damage. You can weave it throughout a slab without requiring a single coupling. It never leaks (unless it's inadvertently punctured), and it does not deteriorate with the chemicals in concrete, nor through electrolysis. This makes it ideal for under-slab water piping, where you can loop an entire system without a single joint.

This buff-colored rigid thermoplastic piping made from CPVC is installed like copper, but using threaded or solvent-welded couplings.

If you're using Wrisbo's AQUAPEX® system, you can even fix a kinked line just by heating the damaged pipe with an electric heat gun. This presents a tremendous advantage over soft copper, which, once kinked, requires a splice or replacement. Likewise, if the concrete crew steps on PEX, it won't be damaged. PEX usually carries a 25-year warranty and a 100-year life expectancy, regardless of how hard or aggressive your water.

Like all plastic pipe, PEX comes in all standard sizes from ¼ in. to 1 in. CTS (copper tube size) and in coils up to 500 ft. long. It can be colored either red or blue, to distinguish hot- from cold-water runs.

Less expensive than copper, PEX also spares plumbers hours of soldering couplings, fitting elbows, and cutting lengths of pipe. A run can go from the basement to a second-story bath in one continuous length of pipe. This source-to-outlet method of plumbing is known as a "home-run" system. But even if you're plumbing in the traditional branched method you'll only install about one-third the number of fittings that would be required with copper.

Where joints are needed, the tube itself is used as a compression fitting. Connections are made by sliding a PEX ring over the tube, and then expanding the tube and ring with a specialized tool. After the coupling or fitting is inserted, the PEX tubing shrinks over the fitting to form a joint so strong it carries a 10-year warranty. Plumbers claim to save approximately 30% to 50% on labor. With greater flow rates than copper, PEX can feed faucets and shower valves using smaller diameter pipe.

Tubing made from PEX comes color-coded in blue and red to simplify layout.

TRADE SECRETS

Another way to save money is to assemble some parts of the plumbing system in advance and then install these components on the job site.

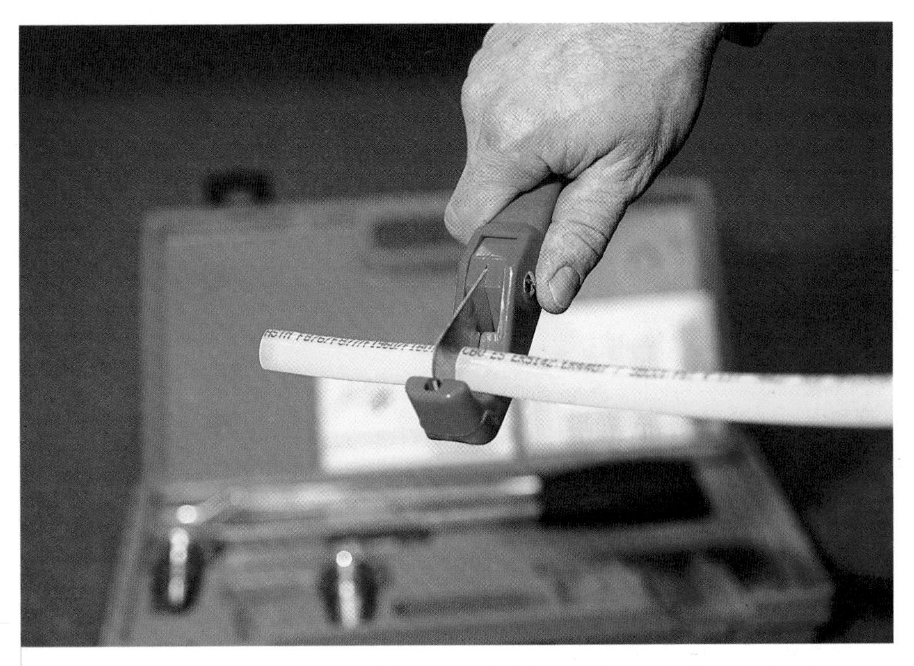
PEX is very flexible, so it bends much like electrical cable. It snakes easily through framing, speeding installation, and long runs can be made without any joints.

In cities that require fire sprinklers, Wrisbo's AQUASAFE® is worth considering. It combines PEX parallel plumbing with fire sprinklers, and using it can eliminate the need for a specialty subcontractor.

Many builders offer the PEX home-run system as an upgrade to copper because it provides practically on-demand hot water, excellent insulating properties, and a reduction in water consumption. But if you install PEX in the traditional branched system, you can use a lot less pipe. Ask your plumber which method yields the lowest cost, but all things being equal, the home-run system is better because it's less prone to leaks. Some plumbers use a combination method, installing a separate manifold per fixture group. PEX is slightly more expensive than CPVC.

High-density polyethylene

There is a less expensive alternative to PEX for cold-water piping: HDP (high-density polyethylene). Although not designed for hot water, it remains highly ductile and resists chemical corrosion. You can encase it in concrete or bury it. Similar to PEX (the same material reformulated to resist heat), HDP has two advantages: It is much more flexible, and it welds to create seamless joints. This means that you string HDP through framing even more easily than PEX, slipping around tight bends and without concern for heating ducts and other obstructions. To join two pieces, the ends are heated with an iron until molten and pressed together. The joint formed is so secure that HDP has become the material of choice for natural gas piping, where no leak can be tolerated.

Using the parallel piping method, every fixture is fed with a "home run" from manifold to faucet.

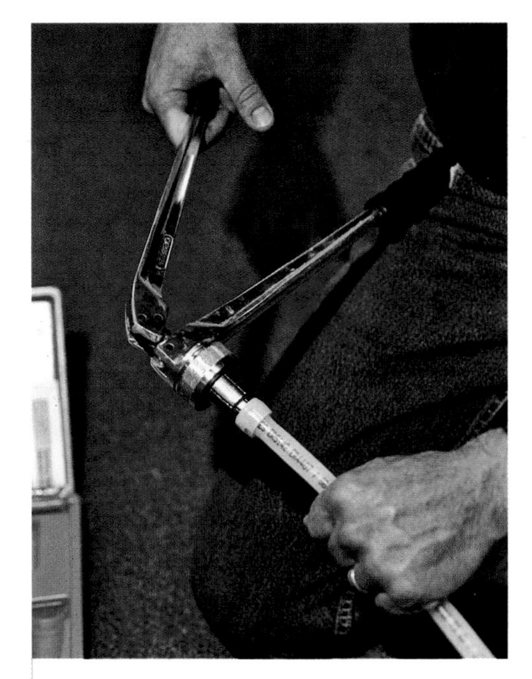

A Vanguard tubing expanding tool expands the end of the tubing so it can be connected to a fitting. A ring of the same material is the compression fitting.

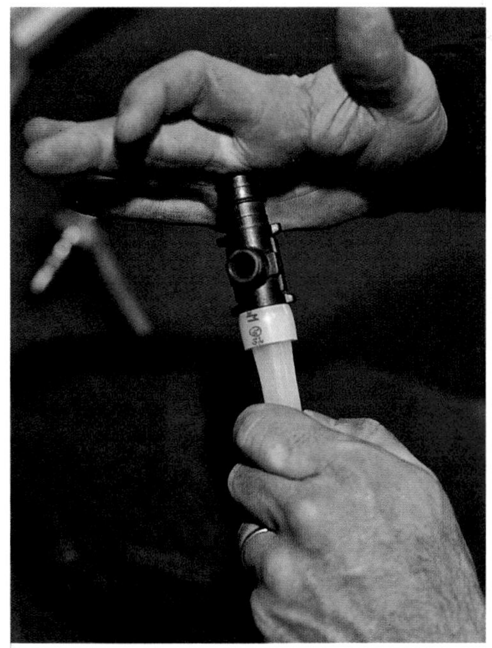

Pressure fittings snap together without solvents to make a strong and very dependable joint.

Eliminating angle stops

Individual shutoff valves at each fixture (called stops) are required by many municipal codes, but not all. If your jurisdiction allows it, consider using a single stop for a group of fixtures. Experience shows that individual angle stops often deteriorate before they are ever used. There is no reason, other than tradition, to have a half-dozen stops in a single bathroom when two (one hot and one cold) would suffice for the whole room.

Another way to avoid using individual fixture stops is to use electronically controlled solenoid mixing valves, which are usually located near the water heater. By blending hot and cold water to the desired temperature at the source, you eliminate the need for dual piping. This system works especially well in conjunction with the PEX home-run system.

THREE WAYS TO PLUMB

In the manifold system, one set of pipes carries water from a cluster of valves to each fixture. In the branched system, pipes branch off to supply fixtures with individual shutoff valves. The combination system (shown) reduces the number of valves and angles and represents the least expensive method.

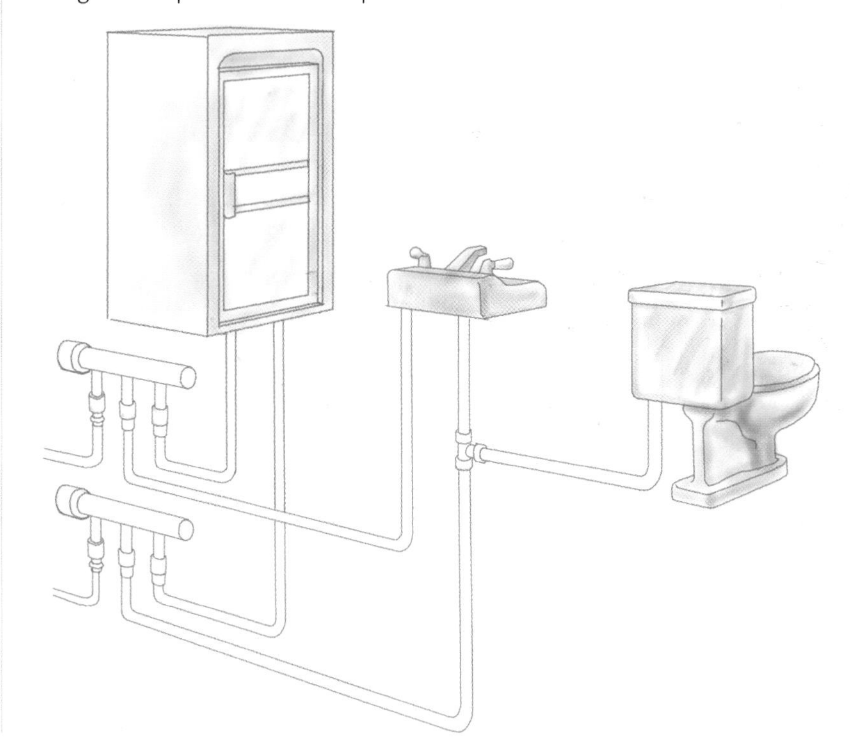

Molten ends of Aquatherm® plastic pipe are joined to create a weld that is as strong as the pipe itself.

BUILDER'S CORNER

Assemble Parts in Advance

Builders who design homes with clustered and stacked fixture groups and build the same plans over and over again can ask their plumbers to assemble prefabricated plumbing "trees," known as service modules, to reduce waste and on-site labor. Some supply houses will build these to a builder's specifications and deliver them on site. Builders who have warehouses can make the space available to their plumbers. The plumbers then can build vertical waste and vent lines, and preassemble tub and shower valves, faucet supply lines, and even the kitchen sink. Prefabrication can take days off a plumber's on-site labor.

Choosing Drain, Waste, and Vent Lines

There are few places in the United States (if any) where plastic sewer pipe is not dominant. But there are cost-saving alternatives even among the most common plastics. No matter what kind of pipe you use, you'll save money by using less of it. This depends primarily on the fixture layout and venting system.

> No matter what kind of pipe you use, you'll save money by using less of it.

Acrylonitrile-butadine-styrene

For general-purpose sewer pipe, the most cost-effective material is ABS (acrylonitrile-butadine-styrene), a lightweight black pipe that cuts easy and welds quickly using inexpensive tools and a one-step chemical solvent. Some jurisdictions prohibit ABS because of its tendency to sag after cycles of expansion and contraction, which makes long pipe runs impractical, and because ABS releases toxic chemicals when it burns. The alternative is PVC.

Polyvinyl chloride

PVC (polyvinyl chloride) can be white, tan, or ivory. It is heavier and more expensive than ABS, and it requires a two-step chemical weld. In some jurisdictions, it's your only alternative. Laminating a sandwich of solid PVC over a cellular core makes a less expansive version known in the trades as "core pipe." It's lighter and easier to work with than conventional PVC and costs about a third less. Because PVC is very stable, there's no need to bed PVC in gravel when installing it in the ground.

Low-pressure polyethylene represents another inexpensive alternative for sewer lines, especially for vertical drops and vents, where you can take advantage of its flexibility to weave it through walls and joist bays.

Eliminating unnecessary vents

Vents provide air in an otherwise closed pipe system. This air equalizes the pressure in the sewer line to keep trap seals from siphoning due to back pressure. The key to reducing overall costs in vent piping is to maximize trap-arm lengths—in other words, to use the longest allowable distance between traps and their vents. Along with reduced venting techniques, this eliminates needless piping.

You can extend a 1½-in. trap arm up to 5 ft. without any problem even though many local building inspectors do not permit distances greater than 2½ ft. Although vent lines usually extend through the roof, there is no reason vents can't terminate through a sidewall or into a soffit vent. This can prevent leaks and allows shingles to be installed before the plumber has finished.

Wet vent

A wet-vent system utilizes a vent for one fixture that also serves as a waste line for another fixture. Whether this approach is possible depends on the type of fixture and the size of the pipe. Using a wet-vent system can reduce total vent lines on a single group of fixtures by more than half.

Stack vent

With stack venting, the only vent is the solid-waste stack (main wet-waste line). No other venting is required. This is by far the least expensive conventional system, although few jurisdictions allow it, especially on multiple-story buildings.

Loop vent

An alternative to the stack-vent system that may be more palatable to your plumbing inspector is the loop-vent method. It is essentially a method of venting an entire fixture group with one line.

Although ABS pipe has decades of reliable service behind it in many areas of the United States, some building departments prohibit it because the plastic produces toxic gases when it burns and because it has a tendency to sag more than PVC.

VENTS THAT DO DOUBLE DUTY

A wet-vent system uses one fixture's vent as the waste drain for another fixture. This approach requires less material than a conventional venting system in which each fixture has its own vent.

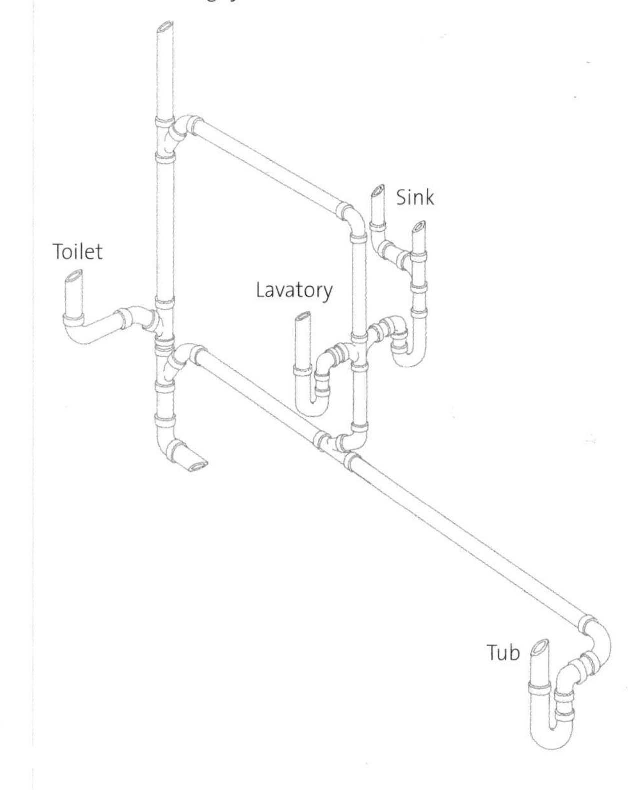

A ONE-VENT SYSTEM

In stack venting, the soil-waste stack is used to vent all of the fixtures in a bathroom.

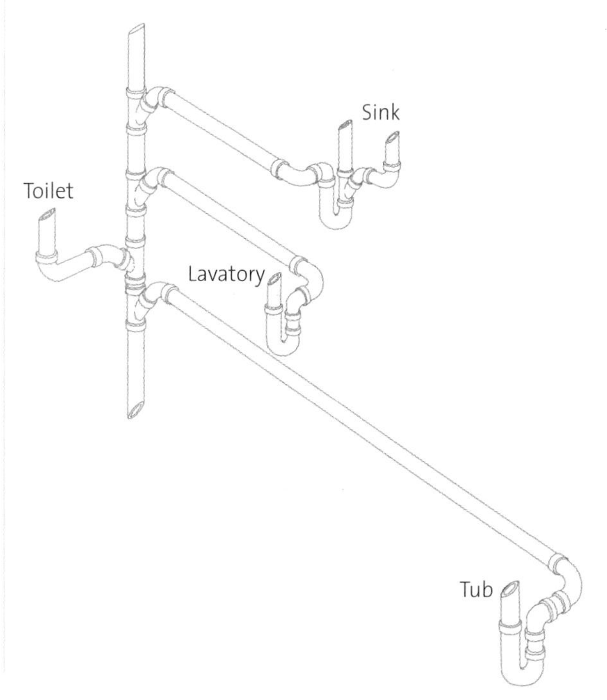

A Second Career

By 1995, Mark Kaufman seemed to have reached the pinnacle of his career. Founder and chairman of the successful Houston-based architectural firm Kaufman Meeks & Partners, he was designing projects for developers throughout the United States and Europe. But Kaufman didn't know his career was only just starting.

Along with his architectural practice, Kaufman dabbled in small real-estate ventures. When he decided to sell an old rental house for $59,000, he had a contract within two hours and wondered why big developers weren't targeting that price range.

He became obsessed with unlocking the key to super-affordable housing, a project that would become his passion and eventually a new career.

He became obsessed with unlocking the key to super-affordable housing, a project that would become his passion and eventually a new career. For three years Kaufman played at the drafting table every free moment, working and reworking floor plans to discover how he could produce a decent home for under $60,000. He realized the solution combined architecture and land planning. He began to lay out high-density neighborhoods with a traditional village feel and found a way to double the standard density by placing eight homes on an acre of land.

Kaufman's homes, ranging from 1,200 sq. ft. to 2,000 sq. ft., sold from $98,000 to $115,000 in 2002. Kaufman designs his homes on a two-foot module and exploits repetitive architectural features. He uses only three window sizes and four kitchen cabinet sizes. He avoids any unusual techniques that might confuse workers and create unforeseen problems.

Since 1998, Kaufman has patented his planning method and built thousands of affordable homes in the Houston area. Under the banner of Parkside Homes, Kaufman's mix of value engineering with neo-colonial architectural and traditional neighborhood standards has proven unstoppable.

Fact Sheet

Who: Mark Kaufman

Where: Houston, Texas

What he did: Saw a market niche for affordable housing and went on to develop land-planning, design, and construction methods that made it possible.

Automatic vents

Costs associated with an open venting system can be avoided altogether by using an air-admittance valve (AAV), or automatic vent. This is an innovative product that greatly simplifies drainage and waste venting by substituting a valve at each fixture group for an elaborate system of pipes. An AAV is designed to open when a plumbing fixture is discharging to allow air to circulate through the system, and then to close by gravity when there is no flow. A positive seal develops once the pressure in the drainage line equalizes so sewer gas can't escape.

Although air-admittance valves appear innovative in many areas of the United States, they have a 30-year history worldwide. These valves meet or exceed all standard performance guidelines and have been approved by all the major codes. Local jurisdictions sometimes limit their use to kitchen island sinks and remodeling applications. Despite the reservations of local building officials, the AAV is not a mechanical or spring-loaded device subject to failure and has a service life estimated at 80 years.

In some areas, the city sewer is actually vented through household systems, and an open vent may be required for this purpose. Using the AAV in combination with a single open vent to satisfy city requirements still saves money. Septic systems also require an open vent that can be installed in combination with these valves.

While most plumbers install one AAV per fixture, such as the Studor® valve shown in the photograph above left, an entire fixture group can be vented just as easily. One Studor minivent is sufficient for two bathrooms. Fixture groups also can be plumbed in circuit-, stack-, and wet-vent combinations. If your jurisdiction allows it, an air-admittance system can save $300 to $500.

This automatic vent manufactured by Studor can eliminate some conventional venting, thus saving money.

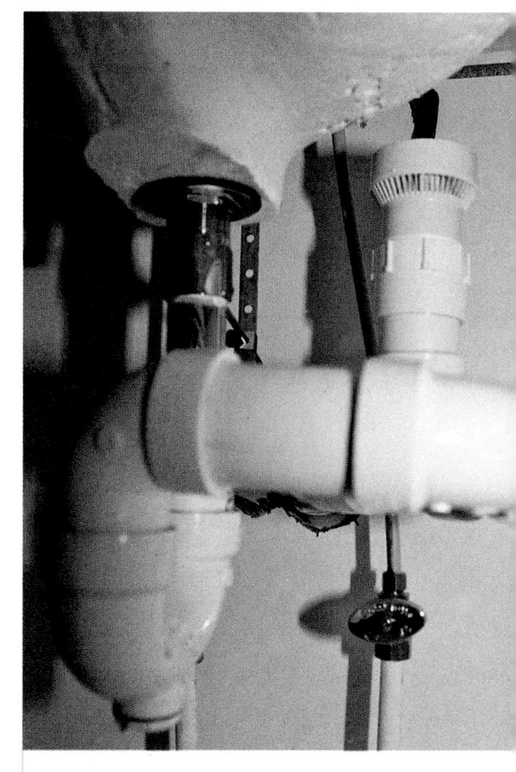

An automatic vent works especially well here, where a kitchen sink is too far away from other plumbing fixtures to share their vents.

Although air-admittance valves appear innovative in many areas of the United States, they have a 30-year history worldwide.

Drain lines can be smaller

In waste pipe, bigger doesn't always mean better. With today's low-discharge plumbing fixtures, small-diameter drains of 1½ in. to 3 in. can work as well as larger-diameter pipe. Larger pipe with the same slope will have lower flow rates, which can promote the deposit of sediment.

Engineers determine the load on a waste system by calculating its drainage fixture unit (d.f.u.) value. This value represents a measure of the probable discharge into the drainage system based on type and total number of fixtures attached. But because it's rare for every fixture in a group to operate simultaneously, you can use drainage lines based on probable use rather than the sum

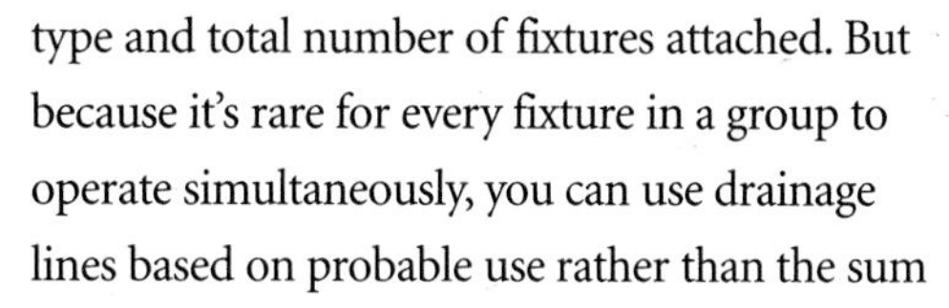

TRADE SECRETS

Sewer gases contribute to sick-building syndrome and give off a foul smell. By preventing the escape of sewer gases into the environment, automatic venting systems not only improve indoor air quality, they reduce one source of damage to the ozone layer.

of all fixtures. As fixtures become more efficient, their discharge is proportionally reduced, dropping the drainage even further.

Unfortunately, plumbing codes have not kept pace. Many building departments require waste lines to be larger than necessary. If you know the actual d.f.u. of the fixtures you're using, you can calculate real values and reduce your pipe requirements. Even using the maximum fixture units from the code table can provide greater economy of pipe than your plumber's intuition. For example, a bathtub and vanity can be drained on a 1½-in.-dia. line without exceeding the horizontal fixture branch requirements of the IRC. Most plumbers would install a 2-in. or larger pipe. With today's power-driven clean-out equipment, below-grade drains can be as small as 1½ in. and cleanouts can be spaced up to 100 ft. apart.

TRADE SECRETS

A handy "slide ruler" provided by the manufacturer helps you determine the minimum CSST diameter required for the length of the gas line and the appliance it's feeding. Just like with any other pipe, right sizing saves money. A ⅜-in. diameter length of CSST costs about 11 percent less than ½ in.

Corrugated stainless-steel tubing resembles flexible electric conduit, but it has a protective yellow polyethylene jacket. Connections are made with ordinary plumbing tools.

Saving with Stainless Steel Gas Tubing

Residential gas-piping options range from traditional black iron to corrugated stainless-steel tubing (CSST). Although more expensive per foot than black iron pipe, CSST is light, easy to install, and, because it's flexible, spares the installer long hours of head-scratching to map the path rigid pipe would require. CSST resembles flexible electric conduit but with a protective yellow polyethylene jacket. Connections are made with ordinary plumbing tools using nuts, washers, and brass couplings. On average, the use of CSST reduces the labor required to gas-pipe a home by 50% to 70% over traditional black iron. This is an odd case where a twofold increase in material costs actually saves money in the overall installation cost.

Seamless flexible copper is also popular in residential gas systems, but it's not as easy to install as CSST. Copper is much more susceptible to corrosion through electrolysis and even chemicals present in natural gas. Flexible copper kinks easily, and since you can't cut and splice it back together, copper becomes more difficult to work with than CSST.

Shopping for Fixtures

You can save the most money in plumbing when selecting fixtures. First, the obvious: White and chrome cost a lot less than color and brass. Just as with groceries, fixture prices vary by wholesaler, fluctuations in seasonal demand, the cost of oil, and, of course, what's on sale at the home-improvement center. But there are some guidelines to follow.

White and chrome cost a lot less than color and brass.

There's a difference between the cheap products you can buy at a discount chain and the competitive builder-grade fixtures the plumbing supply shop carries. It's the kind of difference that might persuade you to pay a little more, but not too much more, for much better quality. The difference is usually invisible to the naked eye; it's inside the fixture. It's worth taking an hour and visiting a local supply shop to see their competitive fixture line.

Tub-and-shower units are easy to install

There's certainly nothing cheaper than a shower pan with acrylic-panel walls and a curtain. But nothing looks cheaper, either.

One-piece gelcoat (fiberglass-reinforced plastic) tub-and-shower units cost less than either one-piece acrylic units or steel and cast-iron tubs with tiled walls. While it's obvious that tiled walls and a cast-iron tub add substantial costs, many builders try to economize further with one-piece shower units. Although a small one-piece shower stall appears less expensive than a tub-and-shower unit, it can have a higher total cost. Shower units usually require extra framing and the installation of a glass door enclosure. On the other hand, a tub-and-shower unit fits snugly between bathroom walls without any furring and requires only a curtain rod to finish.

To avoid the problems associated with fiberglass, which cracks and chips easily during construction, ask your plumber to install an inexpensive tub protector as soon as the unit has been set in place. Make sure to use framed stiffeners (blocks scabbed onto wall framing and up against the walls of the fiberglass unit for support) to reinforce the tub's walls, and stuff some fiberglass insulation around the edges of the tub floor to make it stronger and better insulated.

If you're trying to upgrade and economize simultaneously, consider a steel tub with a simple white-tiled wall. This combination can have a classy look, and if you're in an area where tile work is not expensive, this option can actually cost less than a one-piece acrylic unit. Make sure your plumber uses a mortar bed or foam insulation under the tub to add strength and avoid the cheap-sounding tinny noise you hear when stepping into an unsupported steel tub. Although similar to gelcoat in appearance, acrylic tubs rival cast iron in price. Unless you actually prefer plastic, acrylic tubs cost more and look cheap—just the opposite of what most of us would like to achieve.

If you're trying to upgrade and economize simultaneously, consider a steel tub with a simple white-tiled wall.

China sinks are a good choice

There are three general categories of bathroom sink: cast iron, vitreous china, and porcelain-coated pressed steel. For most applications, vitreous china represents the best choice. Steel is cheaper, but many supply houses don't carry it because it chips and dents too easily. Vitreous china is an inexpensive product

Smaller Can Be Better

The minimum size water line for most fixtures, except dishwashers and lavatories, is 3/8 in. Although many plumbers still install 1/2-in. lines, modern water-saving faucets do not require this much flow to work properly. Lavatories and dishwashers can actually work well with 1/4-in. lines. In most cases, you can feed a single fixture group, such as a bathroom, with a 3/8-in. main. With the exception of dishwasher and icemaker lines, however, plumbers typically use a 3/4-in. main and 1/2-in. branch lines. More volume does not mean higher pressure. In fact, smaller diameter pipe reduces system wear and generally provides higher flow rates and better pressure.

Costing less than a simple 24-in. vanity with a manufactured marble top, this pedestal sink provides an air of luxury to a bottom-dollar bathroom.

that generally makes it through the construction environment unscathed.

Since every manufacturer gets clay from the same Mexican or South American sources and the quality of their products is fairly even, you can shop around for the best deal without too much concern. You'll pay a lot more for a name brand without any discernible benefit except, perhaps, a better warranty. Gerber® makes an excellent line of vitreous china products, including lavatories, which sell for about 15% less than the better-known brands.

Wall-hung lavatories can reduce costs substantially, but most buyers expect a vanity. One alternative is to install a builder-grade pedestal sink, which provides savings and elegance simultaneously. Cultured vanity tops with an integrated basin have become standard fare in tract housing. But these tops are made regionally and fluctuate in price subject to local wage scales and competition. Southern states can get inexpensive vanity tops from Mexico, while in northern regions the same top becomes an expensive upgrade. Where cultured marble is expensive, competitive builders use preformed plastic laminate tops with self-rimming lavatories.

In the kitchen, think stainless steel

Kitchen sinks come in cast iron, manufactured stone, stainless steel, and porcelain-coated pressed steel. Don't even consider a porcelain-coated sink, even though they look almost as good as cast iron. These sinks simply chip too easily. The best alternative is a moderate-quality stainless-steel sink. Some very cheap stainless-steel sinks have shallow bowls and thin-gauge walls that rattle when the garbage disposal is turned on. These are better than the pressed-steel-and-porcelain variety, but not durable. The shiny coating can come right off with steel wool. A longer-lasting, better-quality stainless-steel sink will have the words nickel branding or 302 nickel on the box.

TRADE SECRETS

Studies indicate elevated levels of copper—a corrosion byproduct of copper plumbing—in bathroom and kitchen sinks throughout the United States. Some cities have banned copper for use in residential potable water plumbing. Modern plastic pipes leach no toxic chemicals into even the most corrosive water.

Bargain faucets may really be bargains

Faucets and valves imported from China and Korea have proven both reliable and inexpensive. An imported brass valve often costs the same as a U.S. brand in plastic. Given the choice, brass is better. Chinese and Taiwanese manufacturers take advantage of expired European and U.S. patents to make duplicate products. For example, B&K® faucets sold by the Muller Corporation® are identical to the more expensive Delta® faucets, and B&K faucets can be repaired with Delta parts.

Faucets and valves imported from China and Korea have proven both reliable and inexpensive.

For those of you who resent foreign competition and try to buy all-American, consider the fact that Moen® manufactures its competitive Cleveland™ line in Taiwan, and that a Japanese company owns State Industries, which builds the only water heater still made in America. Nowadays, the reality of international business makes it impossible to buy with a protectionist mentality.

Two-handled faucets and valves are almost always less expensive than those with a single handle. Plastic handles are generally less expensive than porcelain or chrome. So are plastic spouts, but I don't recommend them. It's better to save money by buying a plain-looking but solidly built faucet. This is especially true for tub and shower valves, which may require ripping out drywall or tile if they need to be replaced.

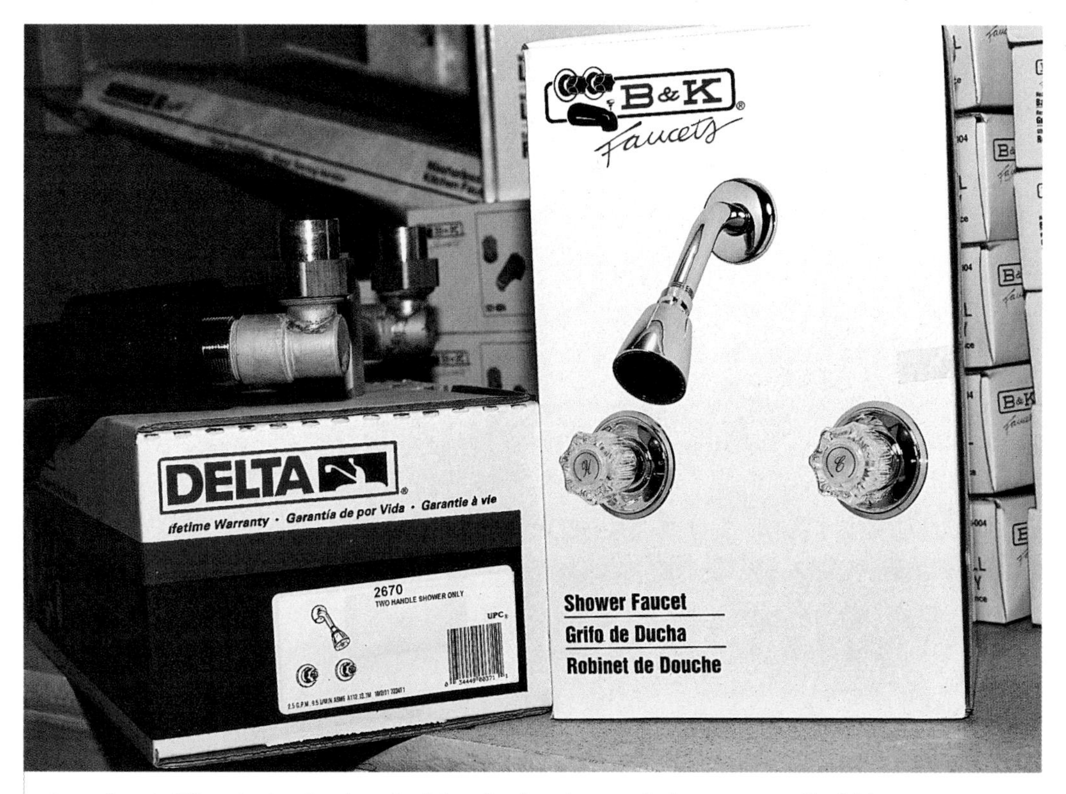

Same faucet, different price: Good-quality "short-line" products make homes more affordable. Manufacturers who only build a limited assortment of styles geared toward competition and not innovation or variety make these products.

Put quality where it counts

I like to save money on lavatory and kitchen faucets, which are easy to replace, but I insist on high-quality brand-name products for anything that's buried inside a wall.

Although most builders don't like to mix and match faucets throughout the house, one money-saving approach is to use brand-name valves where quality counts most, and then use the same manufacturer's competitive line in areas where it's not so important. This saves money while providing a consistent appearance. Both Moen and Delta have good-quality competitive brands. Moen makes the Cleveland line, and Delta makes Peerless®. You can buy parts for either and rely on their guarantees.

> Although toilets look the same, they can be very different on the inside.

If you're smart and do your homework, you can save even more by finding out which brands use third-party manufacturers. The so-called OEM (original equipment manufacturers) sell their products under different brand names. You can buy their fixtures and parts for more money or less, depending on the label—just like blue jeans. For example, Michigan Brass® makes an inexpensive slip-fit tub-shower diverter that costs about four times as much when you buy it under the Moen brand name. Michigan Brass does not advertise and carries no branding costs at all. In this way, you can buy the Moen valve and then buy your trim under the Michigan Brass label and pay less for the exact same product. In the kitchen, I never install a rinse hose because this relatively useless feature adds cost and creates a common warranty complaint.

TRADE SECRETS

If you install sill cocks at each far corner of the house (for example, right front and left back), you can usually serve all four elevations with just two sill cocks without shortchanging your buyer.

Avoid cheap toilets

Because of the do-it-yourself market, you can find many knock-off brand toilets at discount chains. Don't buy one. Although toilets look the same, they can be very different on the inside. The quality of rubber gaskets can

If you don't want callbacks, look for a toilet bowl with internal glazing and good-quality gaskets, such as this Eljer® brand bowl, which is both competitively priced and practical.

A conventional gas water heater requires a flue and gas piping.

In warmer climates, putting the water heater in the attic saves the cost of a long flue duct.

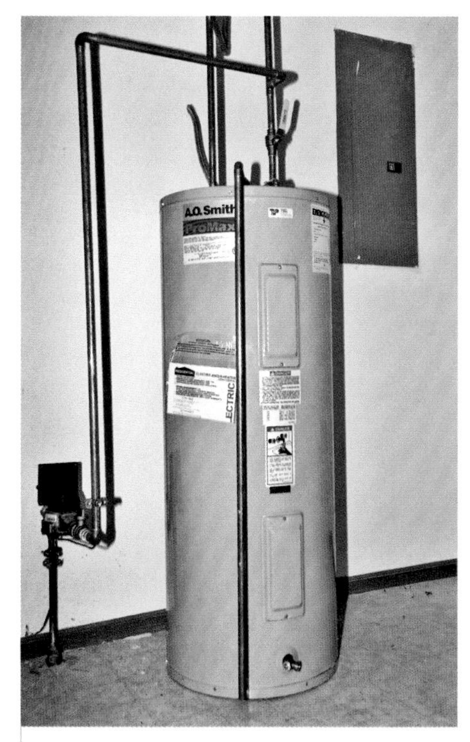

This 50-gal. electric water heater costs about $100 a year to run. Eliminating the water-heater flue required with a gas unit saves about $300.

determine whether you'll be rebuilding that toilet within its first year of use. With only 1½ gal. per flush, glazing in water ports and the trap is very important. Not only will unglazed waterways function more sluggishly, they are prone to plug and are unsanitary.

Gerber makes an excellent, competitive line of good-quality toilets that cost about half as much as a comparable full-line brand, such as Kohler®.

Don't oversize the water heater

A water heater should be sized according to the number of baths and bedrooms in the house. A 30-gal. gas-fired or 40-gal. electric heater will generally be sufficient for a three-bedroom home with two baths. Water pressure can be improved at the faucets and consumption reduced by pairing the water heater with smaller-diameter water pipe and flow-restricting showerheads. This added pressure allows the use of a small heater without compromising comfort.

Although electric water heaters have been maligned for costly operation, today's units are better insulated and much more efficient. Depending on utility costs, an electric water heater might actually cost less to operate than a natural gas unit, while sparing the risk of carbon monoxide poisoning and the expense of running a flue.

A water heater with a 6-year warranty costs about $50 less than one with a 10-year warranty. But this difference in price does not buy you a better or longer-lasting appliance. It simply buys you an insurance policy.

The appliances in this basic kitchen, including a self-cleaning range, a builder-grade dishwasher, and a range hood, cost $900 with tax and delivery.

Saving on Sprinklers

In cities that require fire sprinklers, consider Wrisbo's AQUASAFE system, which combines PEX in parallel arrangement with fire sprinklers. In other words, both cold-water lines and sprinkler lines are combined. As a multipurpose system, your plumber installs the whole thing—potable water and sprinklers—reducing costs by eliminating a specialty subcontractor who would install only the sprinkler system.

TRADE SECRETS

Less expensive dishwashers usually have better energy ratings. This is because they have fewer features and don't consume as much hot water. Although builder-grade dishwashers may not offer "whisper-quiet" operation, they run much more quietly than dishwashers did 10 years ago. To make one even quieter, drape a blanket of water-heater insulation over the tub.

Pick appliances carefully

There's no law that requires every kitchen to have a garbage disposal and a dishwasher. Many builders of affordable homes dispense with both, providing only an electrical outlet and the cabinet space for future installation. Like a garage-door opener and other conveniences, they can be added later.

If you decide to install a garbage disposal, use an In-Sink-Erator® Badger II®, with a ⅓-horsepower motor. It's a recognizable, reliable brand that costs as little as other competitive models. There's no need for an air-gap for your disposal to work properly, even though some local codes require it.

A builder-grade dishwasher is available for less than $200, but this is still a considerable expense by affordable-housing standards. One way to save money is to shop competitive, "short-line" appliance brands, such as Hot Point®. Short-line brands are not low-quality products, but limited lines that don't offer accessories, upgrades, or brand-name recognition. They are manufactured by major companies but not advertised. For example, General Electric® makes Hot Point, Maytag makes Magic Chef®, and Frigidaire® makes Gibson®. There's no reason to avoid any of these less-expensive brands, since they come with the backing of solid, mainstream companies that offer good warranties and provide service centers all over the country.

In general, electric appliances always cost less to rough in than gas appliances do. There is no combustion chamber to vent and no need to run two separate utilities to the same unit. But regional preferences usually dictate the choice. With a gas range or dryer, the gas line should be sized according to the minimum requirements of the unit. In most cases, a free-standing range only requires a ⅜-in. line, although plumbers usually supply it with a ½-in. line. Gas dryers usually require a ¼-in. line. The actual diameter of the supply line depends on the system pressure and length of pipe. But knowing the actual requirement always beats guessing. Guessing generally results in oversizing and wasted resources; at other times, it results in undersizing and compromised safety or function.

Heating, Ventilation, and Air-Conditioning

Since the early 1990s, when the government became involved in limiting energy consumption, the efficiency of heating and cooling equipment has steadily improved. A voluntary labeling program called Energy Star® offered marketing advantages to manufacturers while alerting consumers to what products could save them money in operating costs. Lightbulbs and personal computers were among the first products to bear the Energy Star logo, but the list expanded quickly to include major appliances and, by 1997, even whole houses.

The Energy Star rating system allows an integrated, whole-house approach to energy efficiency rather than reliance on expensive, superefficient air-conditioning and heating equipment. Many builders of entry-level homes still resist the idea. They assume that higher energy

The basic simplicity of a spider- or loop-vent system provides cost savings through reduced ducting and installation labor.

To find out how tightly a house has been constructed, use a blower-door test. Effective air-sealing techniques can reduce the size and cost of air-conditioning and heating equipment.

standards will force them to raise their prices. That doesn't have to be the case. An intelligently designed air-conditioning system coupled with inexpensive insulation and sealing techniques can actually bring costs down.

The reasonable route toward affordable comfort begins with designing your home's space conditioning specifically for your climate, insulating properly, and installing the right windows. Think of a house as a huge container. You could provide indoor comfort either with brutally oversized equipment to make up for drafty construction, or with a Thermos-like exterior that holds a desired temperature easily. Heating and cooling relate closely to design and construction practices, so this section makes a good companion for material elsewhere in the book on framing, insulation, and windows.

> It's tempting to leave the design of heating and cooling equipment solely to an HVAC subcontractor. Don't.

Take an Active Role in Planning

It's tempting to leave the design of heating and cooling equipment solely to an HVAC subcontractor. Don't. The subcontractor may use rules of thumb instead of value engineering to plan the system. It's a haphazard approach that limits the efficiency and performance of the system and costs you money.

Although most of us tend to regard HVAC as beyond our expertise, you can gain enough of an understanding to select the most appropriate and economical equipment for your home and climate. But first you have to understand the basics of HVAC design. The two major components of an HVAC system are the heating and cooling equipment and the ducts. Although we will discuss techniques to optimize both, the first step in HVAC value engineering involves your floor plan.

Engage your HVAC subcontractor early in the design process and make sure your blueprints include a central spot for the air-handling equipment and plenty of room to run ducts. By placing your equipment in a central location, you minimize the length of supply lines and can take advantage of a central return system. Try to provide space for the entire system within the building envelope. Placing the system indoors eliminates the need for mastic-sealed, insulated ducts. This can seem especially challenging in a one-story home built on a slab, but today's appliances are small enough to hide in a closet or an insulated mechanical area in the attic. The 10 sq. ft. of floor space you'll need will pay handsome dividends.

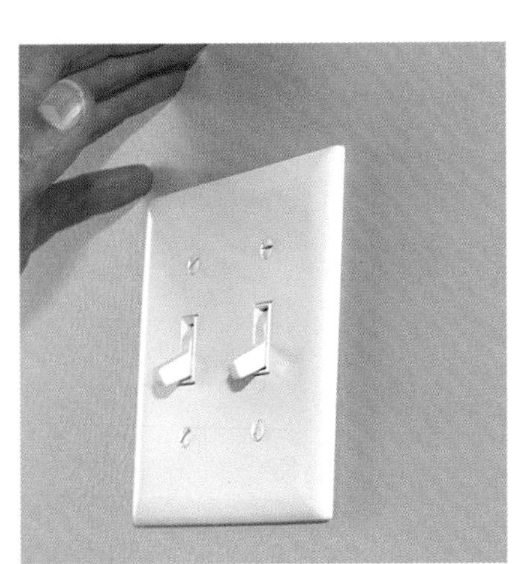

As the blower-door test fan puts negative air pressure on your home, walk around and check for air leaks. You'll be amazed at how much air gets through inconspicuous places such as light switches and plugs.

Lower Costs and Optimize Performance

- Use advanced framing techniques to maximize insulation-to-lumber ratio.
- Install windows with low-e glazing.
- Seal the building envelope to prevent air infiltration.
- Place ductwork and heating and cooling equipment indoors.
- Minimize supply and return ducts.
- Right-size your heating and air-conditioning equipment.

VALUE-ENGINEERED DUCTWORK PLAN

In a value-engineered HVAC plan, place the equipment in a central location to minimize duct lengths and place registers on interior walls.

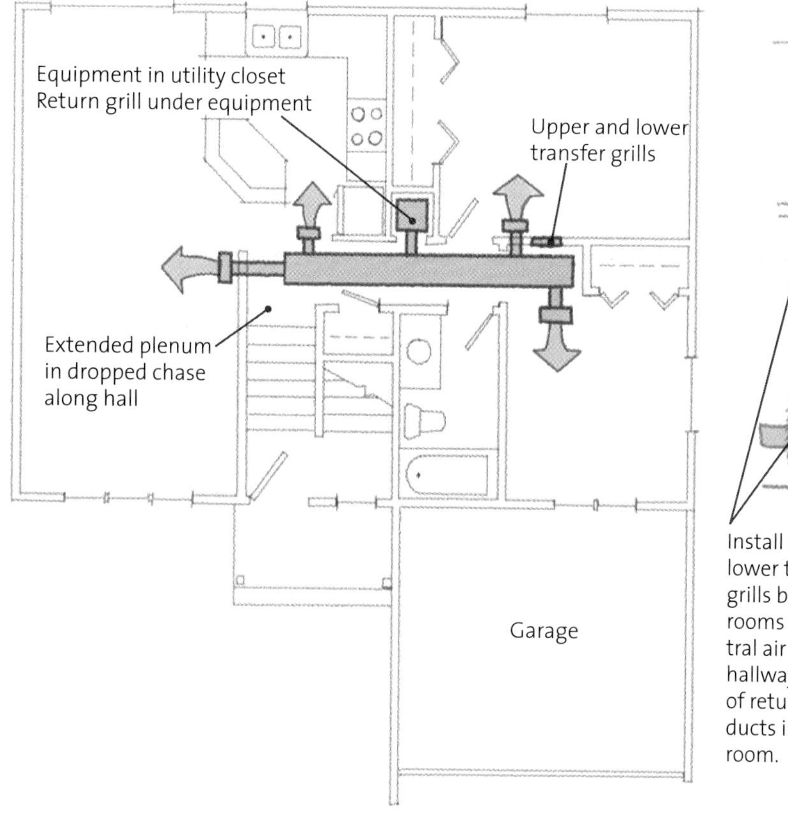

Due to better insulation and humidity control, supply-air registers under windows have become unnecessary. You can install all of your registers on interior walls using shorter, less expensive duct runs. In small, energy-efficient homes, it's not necessary to condition all areas of the house, such as the kitchen, bathrooms, and utility rooms, unless they are large or located on an exterior wall. Where there are no architectural impediments to airflow, one or two registers provide enough air for a large area.

> In small, energy-efficient homes, it's not necessary to condition all areas of the house.

Assuming good construction practices, a well-designed duct system can reduce the cost of air distribution (ducts) by 20% to 30%. Properly sized equipment can reduce the utility bill by another 30% to 50%. By following the full battery of suggestions offered in this chapter, I can reduce HVAC costs in the houses I build by as much as $2,500. My customers cut their annual energy costs by between $350 and $450.

Limit the Number or Size of Windows

Energy-efficient, dual-glazed windows have U-values ranging from 0.35 to 1.25, which translates into R-values of between 1.25 and 2.9. When you compare this to a typical wall assembly that provides R-15, it's easy to see that even the best windows create conduction losses more than 10 times higher than walls.

One way of improving the energy efficiency of a house without resorting to more expensive windows with low-e coatings is to limit windows to 15% of overall wall space. If you're in a high-heat-gain climate (such as the Southwest), consider using architectural shading instead of low-solar-gain glazing. Pay attention to the orientation of your home and place windows to collect winter sun and summer shade.

Just a 90° shift in the orientation of the building can add or subtract one-half ton of air-conditioning capacity. When you can't change the orientation, consider planting deciduous trees on the sunny side of your house.

Insulation and Infiltration

One-third or more of heat loss in a house typically comes from conduction and infiltration through exterior walls and the roof. Because the most significant heat loss and gain is through the roof, builders have gone from R-30 to at least R-38 ceiling insulation in almost all climates.

Walls lose most of their heat through air infiltration. Even an R-15 wall can account for a quarter of the heating and cooling loads on a house due to leaks. In winter, cold and wind force warm indoor air to escape through doors and window joints, electrical outlets, and mechanical penetrations in plates and walls. In summer, infiltration brings humidity indoors, which requires longer air-conditioning run times to remove.

To combat infiltration loads, you can use larger, more expensive equipment or a more cost-effective solution that includes caulking and sealing to tighten the building envelope. The goal is to target key areas of infiltration, such as window and door seals, plate penetrations, the band joist, exterior wall corners, and interior wall intersections. Use caulking, expanding foam sealants, and properly installed air barriers to seal all penetrations through the plates and to the outdoors.

There's no need to make your house super tight, since this triggers the need for mechanical ventilation that defeats your cost-reduction goals. But a reasonably tight house can help lower costs by reducing the size of the air-handling and heating requirements.

TIGHTEN UP, REDUCE COSTS

The size and cost of the heater, air conditioner, and distribution system are substantially affected by the building envelope. Weak points in the envelope, as shown, mean bigger, more efficient, and more expensive equipment must be used.

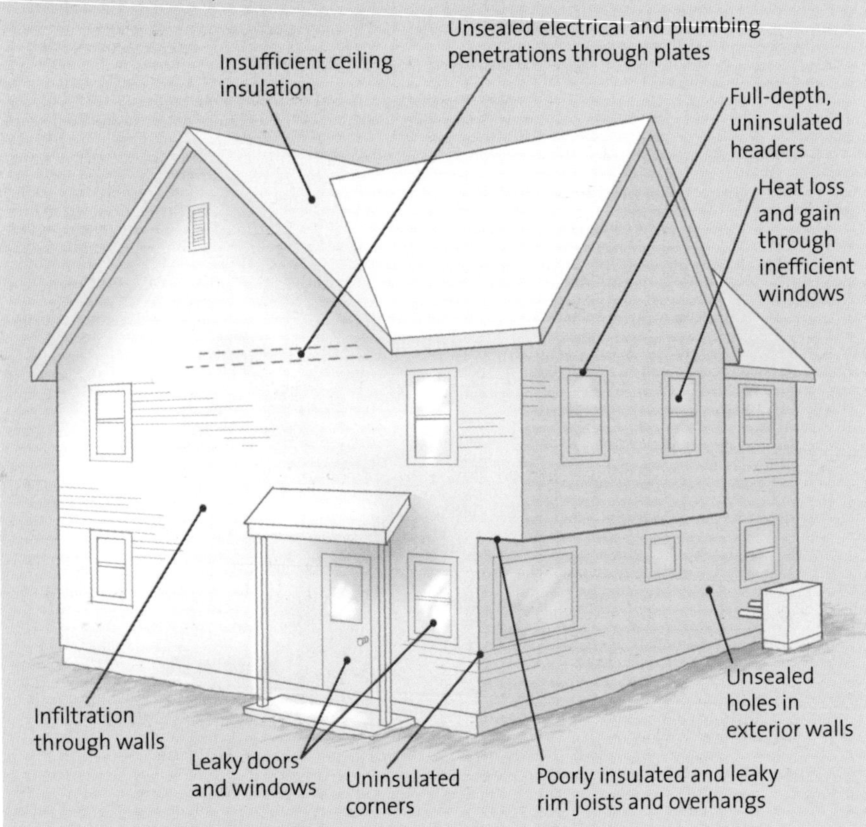

Selecting the Right Equipment

Most builders go with whatever heating and air-conditioning equipment prevails in the area. There are many systems to choose from, including a number of relatively expensive options such as hot-water baseboard, mini-duct systems, hydronic radiant floor, and geothermal equipment. While it pays to look at all the alternatives, forced-air systems are the most economical to install and run. Choosing the right equipment is a two-step process: first the heating equipment, and then the cooling system.

Oil and gas furnaces come in a variety of designs to accommodate different installation and airflow requirements. Whether you choose to put your equipment below, above, or in line with the ductwork has a determining factor on cost.

In general, up-flow furnaces represent the most cost-effective configuration for almost any central-heating application. As its name implies, the up-flow furnace is paired with ductwork installed over the equipment. You'll frequently find this type of furnace standing in a basement. But even where there's no basement, you can use an up-flow furnace and run ducts through a truss ceiling, dropped chase, or attic. A slightly more expensive variety called a "lowboy" is designed for a crawl space or other tight locations.

The down-flow furnace costs more, but you can use one over a crawl space or where ducts run under the slab. Down-flow furnaces come in roof-mounted package units, too, but here the consolidation of equipment offsets their higher price (see "Package Units Are Inexpensive to Install" on p. 126), so—in warmer climates—this is often a very affordable alternative. The downside is running ducts through the attic, which doesn't add to the installation costs substantially, but certainly adds to operating costs. The same occurs with attic horizontal units.

Up-flow furnaces are usually installed in the basement, where supply ductwork will run above the unit. These furnaces come in a greater variety of heating sizes and at a lower cost than other types.

Horizontal furnaces typically go in the attic, where the distribution system runs parallel to the air handler. They cost more than other units.

Horizontal furnaces are the most expensive variety, used primarily for attic and crawlspace applications. A horizontal furnace placed in the attic costs more, and you'll need a bigger and more efficient unit to offset energy losses that come with placing your furnace in an unconditioned space. Although you can save floor space by placing your furnace on the roof or in the attic, unless you live in a warm climate, you'll pay dearly for that extra 10 sq. ft. of floor space you've gained.

> In general, up-flow furnaces represent the most cost-effective configuration for almost any central-heating application.

Electric furnaces are efficient

Although natural-gas furnaces are the most popular variety, electric furnaces have a lower initial cost and don't require gas piping or venting. They run at 100% efficiency and may actually cost less to operate than a gas appliance in certain areas of the country. That's especially true when they are combined with a heat pump (see "Heat Pumps Do Double Duty" on p. 125).

An electric furnace is essentially an air handler with resistance heating elements inside. Since an electric furnace is compact and requires neither venting nor combustion air, it can go virtually anywhere in the house. Sizes can be confusing at first because electric-heat ratings are given in kilowatts instead of British thermal units (Btu). Each kilowatt equals 3,412 Btu. These furnaces can accommodate any airflow configuration and have become increasingly popular, even in extreme heating climates such as the upper Midwest.

Split systems typical for cooling

Today, most residential cooling is supplied by a "split system," which consists of an outdoor condensing coil and compressor connected to an indoor evaporative coil with refrigerant

In a smaller home, you can cut the cost of the furnace by one-third by using a water heater as a heat source. This water heater circulates hot water through a coil in the air handler and a radiant floor system. Without radiant-floor heat, a 50-gal. tank works for most installations.

Research for Builders

Every industry needs research and development to grow. Few homebuilders can afford to explore new technologies and products on their own, but for more than 40 years the National Association of Home Builders Research Center has been doing just that.

In a subdivision not far from their offices in Bowie, Maryland, the Research Center recently built four homes using designs, materials and methods that will allow builders to deliver high quality, affordable homes with broad market appeal. Called MADE homes (for Marketable, Affordable, Durable, Entry-Level), these houses include precast foundation walls, webbed floor trusses, advanced framing techniques, cement- board siding, innovative plumbing systems and standing-seam metal roofs.

Just as all demonstration homes at the National Research Home Park, the four MADE homes remain open for one year so that builders can visit and learn about the techniques and products used. After this, the homes go for sale. Even after they're occupied, the Research Center continues to monitor the houses to gather information on their long-term performance.

> The National Association of Homebuilders Research Center in Bowie, Maryland provides state-of-the-art educational facilities and residential construction testing laboratories.

In addition to the Research Home Park, the center has a state-of-the-art educational facility and testing laboratory that studies everything from the structural capacity of new earthquake and hurricane designs to the flush-performance of hundreds models and brands of toilets.

Anyone interested in innovative building practices will find a treasure trove of information in the Research Center's publications and at its website (www.nahbrc.org). The Research Center also provides a hotline service called the ToolBase Program that answers questions on any aspect of residential construction, either exotic or mundane, within 24 hours. This service is available at no charge to builders, remodelers and consumers by calling (800) 898-2842 or emailing toolbase@nahbrc.org.

Fact Sheet

WHO: National Association of Home Builders Research Center

WHERE: Bowie, Md.

WHAT IT DOES: Explores new techniques and materials that help builders provide affordable, high quality homes.

tubing. As the furnace blower moves air across the evaporative coil, the circulating refrigerant dehumidifies and chills the air. Mechanical engineers measure the nominal size of an air-conditioning unit in "tons," each of which represents 12,000 Btu. Residential equipment ranges from 1½ to 5 tons. A unit's SEER (seasonal energy-efficiency ratio) represents its energy efficiency—the higher the number, the more efficient the equipment. Most units range from 10 to 12 SEER, with higher efficiencies available at a premium.

In dry regions, an inexpensive evaporative cooler (also called a swamp cooler) can provide air-conditioning by drawing outdoor air through a moistened filter in a fan cabinet located on the roof. Although these units cost a lot less than a refrigeration system, the roof-mounted cabinets need frequent cleaning and adjustment. For the cooler to work properly, windows have to remain open, creating security concerns and erratic comfort levels. Air movement provided by ceiling fans or a whole-house fan can contribute to indoor comfort, but these fixtures generally support rather than replace a central air-conditioning system.

Evaporative coolers add to indoor humidity levels and work best in desert climates. They cost about one-third as much as standard air-conditioning systems.

Most residential split systems include an outdoor condensing coil and compressor.

Heat pumps do double duty

A heat pump heats and cools by reversing the direction of the refrigerant in the system. Although heat pumps cost a few hundred dollars more than a standard air conditioner, they can save money when used in temperate climates—where temperatures don't dip below 30°F—because they both heat and cool.

In colder areas, a heat pump works well for most of the year, but it may need an electric or gas furnace backup on very cold days. In conjunction with a standard-efficiency furnace, this kind of system provides low energy consumption on an annualized basis. It also costs less than a typical split system with a high-efficiency furnace, and so it's a good way to get a high-efficiency system at a lower initial cost.

A heat pump uses ambient air to heat or cool. In mild, mixed climates, a heat pump can provide winter and summer comfort. In cooler climates, a heat pump requires a gas or electric backup heat source.

TRADE SECRETS

A raised equipment platform under an air handler in a utility closet can provide a central return grill and filter rack for most small single-story homes.

Self-contained units, sometimes called "package units," combine the blower, coils, condenser, and heating elements in one cabinet. The equipment can be roof mounted or located at ground level next to the house.

Package units are inexpensive to install

The most economical means of providing air-conditioning and heat is a self-contained or "package" unit. It includes a compressor, a blower, and all the coils in a single appliance along with a gas or electric heater. These units can be set on a concrete pad on grade and ducted through a crawl space or, more typically, on the roof with an insulated attic duct system.

These units work best in climates where cooling is a priority, but you'll find them on commercial rooftops in all areas of the country. They cost less to install than other kinds of equipment, but they also cost more to run.

TRADE SECRETS

Oversized air conditioners, besides inflating purchase costs, reduce energy efficiency, lower humidity control, and shorten product life due to excessive on-off cycling.

Sizing Equipment for the House

The price of heating and cooling equipment is a function of its size and efficiency. Calculating the correct capacity of the equipment is more than guesswork. Climatic conditions, the cardinal orientation of your home, and building practices all influence the decision. An HVAC contractor or an outside consultant typically will prepare heat gain and loss calculations to determine how big the equipment should be—but if you don't get involved, these calculations may not reflect your construction methods accurately.

Equipment sizing software comes from a variety of organizations, including the American Society of Heating, Refrigeration and Air-Conditioning Engineers (ASHRAE) and the widely used *Manual J: Load Calculation for Residential Winter and Summer Air Conditioning* from the Air Conditioning Contractors of America (ACCA).

The HVAC contractor or an outside consultant uses a local climate modifier, along with details from your plans. If this data does not reflect your construction practices accurately, you may miss an opportunity to optimize your equipment. Every effort you make to upgrade insulation and air-sealing has an inverse effect on the equipment, allowing you to install smaller, less costly appliances. Overseeing the data input ensures that you receive full credit for your efforts.

The price of heating and cooling equipment is a function of its size and efficiency.

You'll also have to guard against a tendency in the HVAC industry to oversize equipment, especially air conditioners. Some installers believe that if certain size equipment is adequate, then the next size up works even better. The opposite is true. Equipment based on right-sized heat loss and gain calculations always works more efficiently and provides greater comfort. If you have to decide between rounding up or down on a given calculation; round down. Every 10,000 Btu or half-ton cutback in your equipment translates into hundreds of dollars in savings as well as better performance.

Calculating heat gain and loss

Heating calculations entail a room-by-room analysis to determine the amount of heat and airflow needed to keep all the areas of your home comfortable. Calculations reflect the heat transferred through walls, windows, doors, ceilings, and the foundation. They also take into account how leaky your home is and how well you insulate and seal your ductwork. It makes more economic sense to seal your house tightly and enclose ductwork in the conditioned building envelope than it does to pay a penalty in larger, more costly, and less efficient equipment.

> It makes more economic sense to seal your house tightly and enclose ductwork in the conditioned building envelope than it does to pay a penalty in larger, more costly, and less efficient equipment.

Just like the calculations for heating, your cooling computations require a room-by-room analysis. But cooling calculations become considerably more complicated because in addition to heat gains through the building envelope (similar to heat loss), you have to take into account heat gain from occupants, lighting, household appliances, and solar radiation through the windows. And, as if this wasn't enough, the additional energy required to remove excess humidity from the air is a significant increase for overall cooling loads, ranging from 25% to 30%.

The software used to calculate cooling-unit size computes local climactic conditions and the quality of the building envelope and arrives at a number of tons of cooling capacity. It's important not to exceed the required tonnage because oversized equipment operates in short bursts, cycling too quickly to bring the humidity level in the house down to a healthy and comfortable level. Oversizing by as little as 25% can cause considerable problems in a humid climate. Frequent cycling also wears out the equipment prematurely and devours energy.

Getting the right balance Because the requirements for heating and cooling differ, a mixed system has to balance the airflow requirements of both. To ensure proper balance, specify indoor and outdoor design temperatures to the HVAC contractor. Typically, indoor winter and summer temperatures are 68°F and 76°F respectively. You'll find nominal recommended outdoor temperatures in many published guidelines or at your local building department.

The ACCA recommends a maximum of 40% oversizing of gas-fired furnaces when coupled with air-conditioning coils and a maximum of 25% oversizing for air conditioners and heat pumps. Conservation experts recommend guidelines that are even more stringent, suggesting a slight undersize instead. Make sure to specify your preferences since it's common practice to specify the largest equipment allowed.

TRADE SECRETS

You don't need to provide extra humidity during winter months because improved air-sealing techniques preserve indoor humidity to the point that year-round dehumidification is more appropriate.

Planning the Network of Ducts

An air-distribution system consists of supply and return plenums, trunk lines, ductwork, and registers. Plenums collect air and distribute it to and from trunk lines that, in turn, collect air from individual supply and return ducts. As their name implies, supply ducts carry conditioned air throughout the house; return ducts carry it back into the system for filtering and reconditioning. Registers are the grills that hide duct terminations from view; supply registers also control the direction and throw of air into a room through fixed or movable louvers.

After choosing the type and size of air-conditioning equipment, you and your HVAC contractor can lay out a duct plan. Ducting should never come as an afterthought, just as you would never decide beam sizes or shear-wall nailing on the fly. Just as you can use software to size equipment, the ACCA's *Manual D: Duct Design for Residential Heating and Cooling* can help calculate duct requirements room by room and outlet by outlet. Undersized ducts accelerate and squeeze the air, creating noise and leading to higher operating costs. Oversized ducts cost more and result in system imbalances.

> Ducting should never come as an afterthought, just as you would never decide beam sizes or shear-wall nailing on the fly.

Once the requirements for every supply are known, a duct layout can be designed to provide optimal air mixing based on the demands of the climate. The type of duct system you use and how you lay it out depends on the type home of home you're building and whether you build in a predominantly heating or cooling

TRADE SECRETS

Leaky ductwork is responsible for substantial energy losses and comfort problems in most residences. In the attic, garage, or crawl space, seal your ducts and equipment properly with aluminum-backed tape, mastic, or butyl-backed tape.

EXTENDED PLENUM SYSTEM

Typical distribution system components for an extended plenum system: One of the most common layouts in basement construction, which is also useful when a trunk line runs along an interior hall.

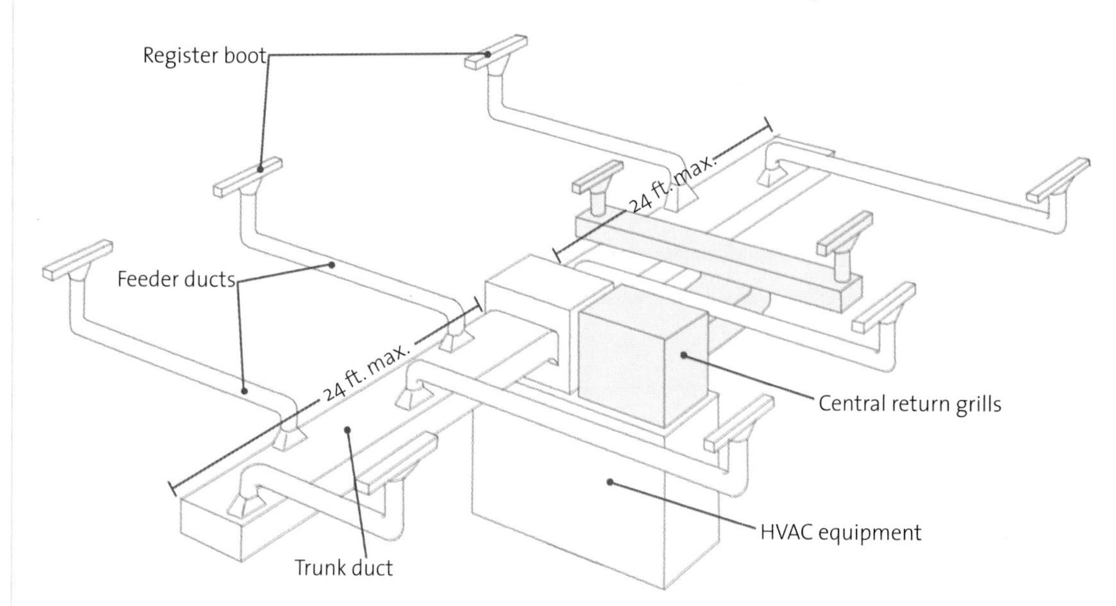

Duct Efficiency Lowers Costs

The overall efficiency of a heating and cooling system is a function of both the equipment and the ductwork. Bringing ductwork inside the building envelope may allow the use of standard-efficiency equipment. In the end, that's a less expensive route than buying high-efficiency equipment and leaving ductwork outside a conditioned space.

Case comparison	Equipment efficiency (in %)	Duct efficiency (in %)	Systems efficiency (in %)	Cost considerations
Case 1: High-efficiency gas-heating equipment and all ducts in unconditioned space	92	70	64	Cost of high-efficiency equipment vs. standard-efficiency equipment
Case 2: Standard-efficiency equipment and all ducts in conditioned space	78	98	76	Cost of locating ducts indoors, minus the cost of smaller ducts, no duct insulation, less stringent duct sealing, and smaller equipment

Adapted from *A Builder's Guide to Residential HVAC Systems*, NAHB Research Center (Washington, D.C.: Home Builder Press, 1997.)

TRADE SECRETS

The least expensive heating and cooling system is a "package unit," but they are more expensive to operate.

environment. But three general guidelines for cost cutting apply to any floor plan and climate:

> Keep ducts and equipment in the conditioned building envelope and avoid attics and exterior walls.

> Install the simplest design with as few joints and bends as possible.

> Locate outlets on interior walls, close to the central supply, and use a central return with transfer grills instead of multiple returns.

Use sheet metal supply ducts in conditioned spaces

Supply duct materials range from cold-formed galvanized sheet metal to fiberglass duct boards. In homes with basements or crawl spaces, sheet metal remains the preferred choice. It's inexpensive, strong, and easy to use. Slab homes with attic distribution systems generally have flexible, insulated ducts or fiberglass duct board. These materials cost a little more, but they require less labor to install. Each material has its advantages. The choice depends on where ductwork is located and what distribution system you choose.

A register boot tied to a central plenum provides a cost-effective means of providing air distribution on an interior wall.

RADIAL SYSTEM

Typical distribution system components for a radial system: One of the most common layouts in crawl-space construction, which is also used with PVC ducts under a slab.

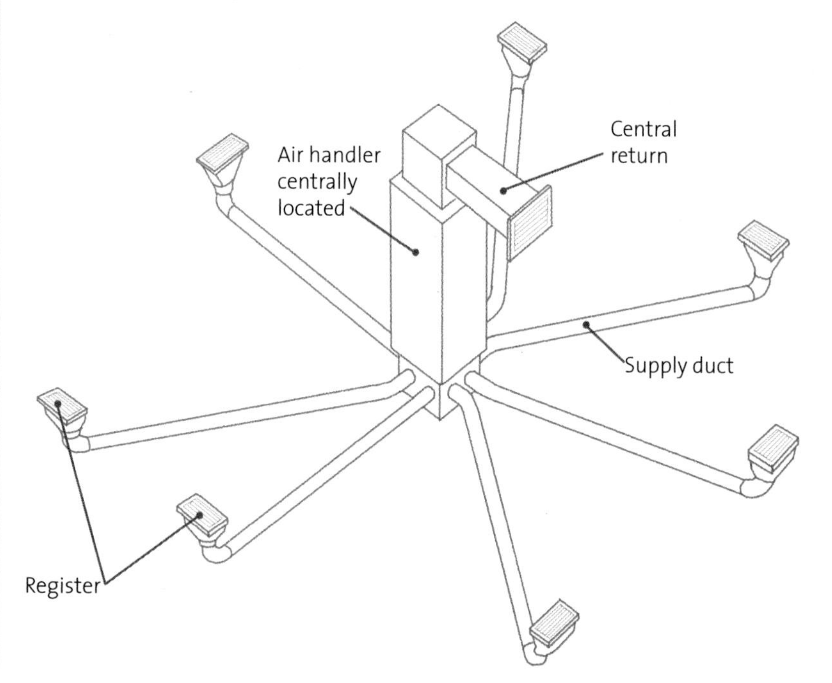

In the basement, use an extended plenum system Basements provide ample room for an efficient distribution system—and it will be contained in conditioned space. The extended plenum method is the most common and effective approach. A large main-supply duct connected directly to a plenum at the furnace serves as a plenum extension; individual supply ducts branch off this central trunk line to individual outlets.

> The extended plenum system is a low-cost way of distributing air in basement, bi-level, and split-level homes.

To take full advantage of this system, locate the heating and cooling equipment near the center of the house and T the trunk line directly off the supply plenum at the furnace—supply lines and branch ducts will run in both directions from this central point. If you can't locate the furnace centrally, put it in-line at one end of the trunk, but keep in mind that an extended plenum can only run about 24 ft. before it loses air pressure.

Locate all registers directly off or right above the central trunk to avoid extended duct runs. I often install register boots right onto the central trunk line and supply the entire house without any branch lines at all.

The extended plenum system is a low-cost way of distributing air in basement, bi-level, and split-level homes. You can use a similar approach in slab-built homes by hiding the trunk line in a dropped ceiling, such as in a central hallway.

Radial systems are hard to conceal A traditional approach to crawl-space distribution involves a centrally located downdraft furnace with a radial duct system underneath. This method also applies in homes built on slabs with plastic ducts installed under the concrete.

A spiderweb-like distribution network of supply ducts running straight from the furnace to each outlet provides a direct method of air delivery that maintains balanced pressure throughout the system. This provides the most economical distribution system for many single-story applications, especially when coupled with a centrally located supply register that helps shorten supply runs. However, the radial system doesn't work well if you need to conceal the ductwork—such as in a finished basement—so outside of attics and crawl spaces, it has limited practical application.

Radial systems in the attic You can also install a radial system in the attic with an up-flow or vertical-vent furnace. The disadvantage of this system comes with placing ductwork in an unconditioned space. No mater how well

TRADE SECRETS

Only a handful of manufacturers make all the HVAC equipment available, and they sell the same models under different brand names at various price points. For example, Ruud makes a budget line called WeatherKing. It's the same equipment as they sell under the Ruud trademark, but without the popular name recognition. If your dealer can provide you this off-brand, you can obtain high-quality equipment for about 40 percent less.

you insulate and seal the ducts, in an unconditioned area they never perform as well as they would inside the building envelope. Some builders use reflective roof sheathing or even fully insulated rafters to provide some insulation, but it's always less costly and more effective to bring the ductwork indoors.

One way of getting around this problem is to build a knee wall in the attic and insulate it. It's easy to condition this small space by splicing a register right into a supply duct. The knee wall system works well for small homes, but it loses its cost-effectiveness when you have to install a knee wall on both sides of the attic to reach every distribution point.

Dropped ceilings and the plenum truss Because of their web design, trusses can feel like a labyrinth for an air-conditioning contractor trying to install the ductwork. In order to keep ductwork overhead and yet enclosed within the building envelope, you can either install ducts in bulkheads over discreet areas, such as closets, or use a dropped ceiling in a central hall.

An innovative homebuilder in Florida designed a new type of truss with a built-in plenum space that works to provide a large, free distribution area within the building envelope, without sacrificing any ceiling space.

Floor-truss plenum has drawbacks Although it's inherently inefficient, a pressurized floor-truss plenum provides the least expensive distribution method available. Conditioned air is pumped into the sealed envelope of a floor-truss system. Outlets are cut into floors and ceilings wherever a register is needed. Builders once installed a similar system by pressurizing a sealed crawl space. These systems work well in theory, but it becomes impractical to seal and insulate a floor or crawl space tightly enough to contain pressurized air without excessive heat loss and infiltration.

The basic simplicity of a spider- or loop-vent system provides cost savings through reduced ducting and installation labor.

A Word about Registers and Diffusers

GRILLS Grills are air inlets with louvers and no damper system. They conceal the duct opening at returns and provide for passive air movement between rooms. Although they cost little and it's tempting to use them in place of diffusers, grills don't throw air into the room and won't provide the kind of mixing required for comfort. In a central return system, you can use a hinged return air face with a filter rack for easy air-filter replacement.

REGISTERS Registers come equipped with louvers and dampers. They generally are installed in the floor, in the baseboard, or on the wall near the ceiling and distribute air in a fan-shaped pattern. Designed to blanket exterior walls with conditioned air, the older registers had a single damper. Although slightly more expensive, registers with opposed-blade dampers provide more uniform air distribution than single-blade dampers. Registers work best in predominantly heating climates. When using an interior-wall register system, registers should be located on the wall about 6 in. from the ceiling for optimal air mixing in warm climates, on the floor for predominantly heating climates.

DIFFUSERS Diffusers deliver air parallel to the adjacent surface. You'll find them mostly in ceiling applications, where they provide superior air distribution in warm climates. Some diffusers have adjustable louvers, which allow the airflow pattern to be adjusted. In the heating season, air can be directed downward for better air movement.

With simple framing, ductwork can be concealed in a bulkhead or chase.

Wall- and floor-cavity distribution

Pressurizing a stud or joist bay also can form ducts. While you'll see this method used for return lines, building departments and HVAC contractors shun wall-cavity supplies because condensation within the cavity can ruin drywall and promote mold. But in dry, desert climates, where humidity is not a problem, this method provides an inexpensive alternative.

The key to using it successfully is to seal the drywall to the studs with either caulk or a vinyl gasket. Some builders fashion a duct liner by stapling polyethylene to the inside of the stud bay before drywall is applied. Others use sheet metal and fiberglass liners.

Although cavity vents don't cost much to build, they tend to leak pressurized air and increase the operating budget of a home. Good sealing techniques can offset losses, but you have to maintain high quality control over several stages of construction to achieve it. It's common to use this system in return ventilation because the pressure is relatively low. But in either case, you should never install a cavity vent on an exterior wall or a floor

PLENUM TRUSS BOOSTS EFFICIENCY

A special kind of roof truss developed in Florida includes a raised center section where ductwork can be concealed. Keeping ducts inside the building envelope reduces demands on the heating and cooling system, meaning greater efficiency and lower costs.

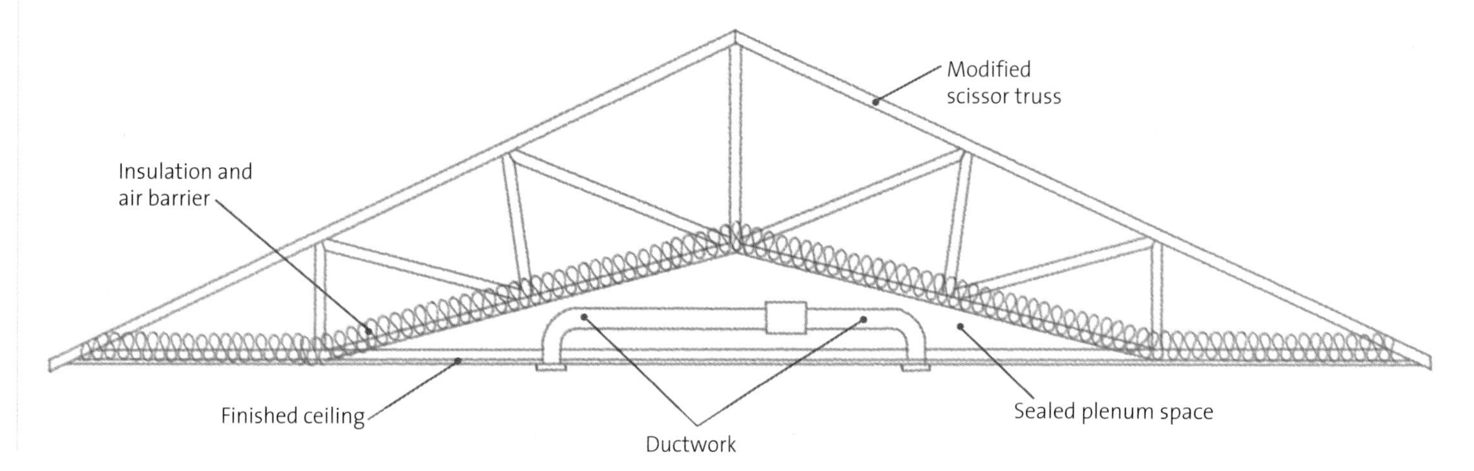

BENEFITS OF A PASSIVE AIR SYSTEM

A passive ventilation system is effective in moderate-sized homes with an open floor plan. Because it requires only one large, relatively short duct, it is the most economical system available and may allow the use of a smaller blower.

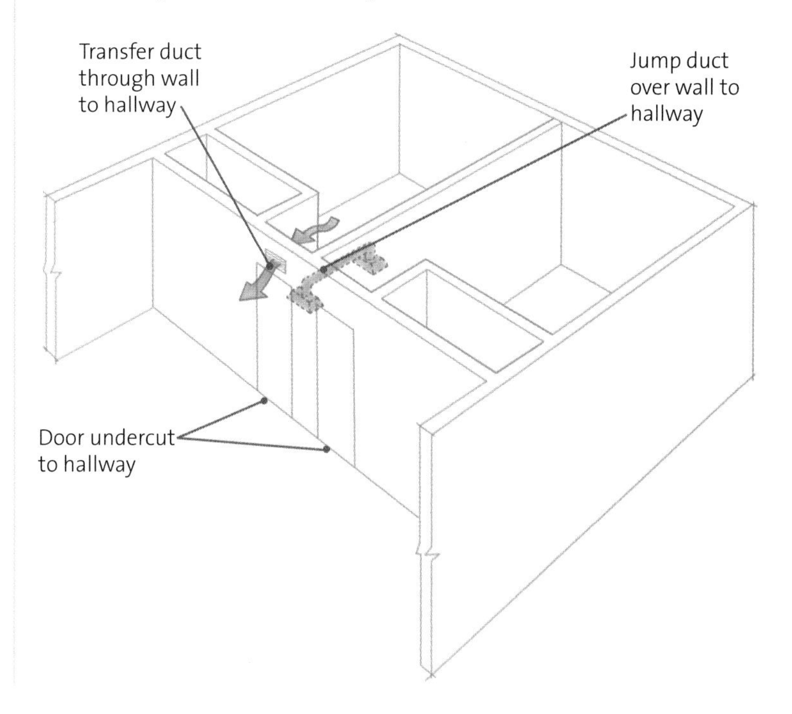

Money-Saving "Green" Checklist for HVAC

- Locate the mechanical room centrally to reduce the amount of ductwork required.
- Install ductwork within conditioned space.
- Eliminate fireplaces.
- Install the correct-size air-conditioning equipment.

A flexible canvas connector helps dampen vibration noise in a central return system.

cavity that extends out to a rim joist. While indoor leaks diminish your control over the delivery of conditioned air, exterior leaks break your building envelope.

Systems for returning air

The air-return system shapes airflow patterns within your home to ensure the air supply is mixed thoroughly before it's recycled for filtering and reconditioning. Good balance between the supply and return systems guarantees optimal performance.

Returns can work either actively or passively. Active returns are connected to the air handler through ductwork. Passive vents aid the movement of air through the house by means of transfer grills, door undercuts, and jump ducts (short ducts that pass over walls). Passive returns cost little or nothing to install while providing an excellent means of blending household air.

Passive returns cost little or nothing to install while providing an excellent means of blending household air.

Passive return systems are low in cost A central return consists of a single active vent ducted in close proximity to the air handler. It draws air from a central location, such as the living room or a hall, through one or more grills. In multistory homes, grills can be stacked in a common wall cavity or duct chase. These grills work in conjunction with a passive ventilation system to circulate air. This approach works well in moderate-sized homes with open floor plans. It's certainly the lowest cost system available because it

PAN VENTS CAN REDUCE COSTS

To make multiple-room return systems more affordable, some builders turn wall and floor cavities into ducts. The downside is a higher potential for leaks and system imbalances.

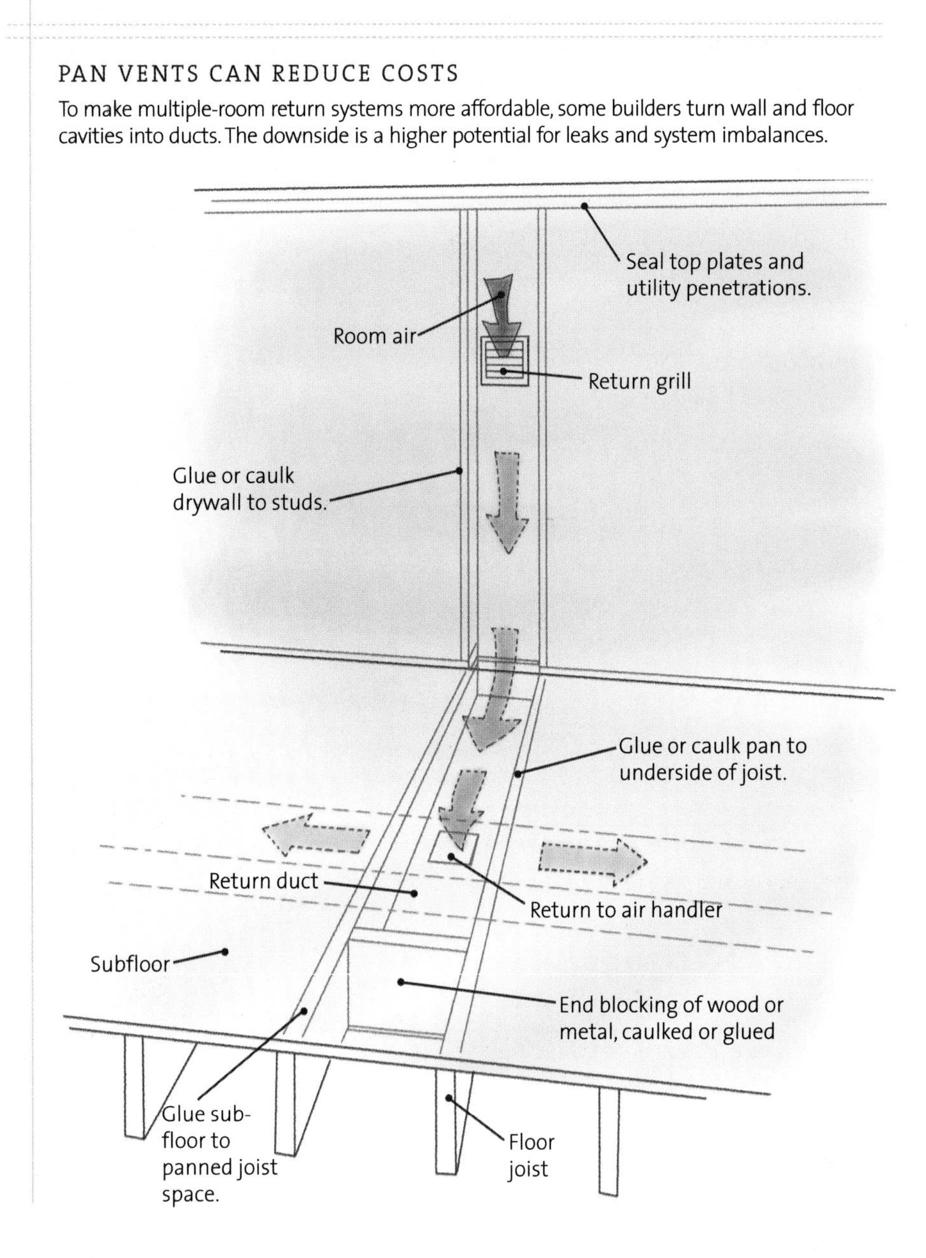

requires only one large duct of relatively short length. With so little ductwork, there's almost no friction in this system, so the blower can be smaller.

Some HVAC contractors object to a central return system because it can be noisy. This usually results from an oversized air handler that produces air turbulence in the lines and transmits mechanical clatter. Besides sizing your equipment correctly, you can attenuate noise by installing a flexible connector between the return plenum and trunk line and return register boots with a curved, rather than angular, throat to provide smoother airflow.

Multiple-room returns recycle air from each room. Although this may seem desirable, it often results in an unbalanced system—some rooms having too little pressure and others having too much. But multiple-room returns generally run more quietly and afford more privacy. If designed correctly, a multiple-room system provides better airflow within each room, even with the door closed. Of course, multiple-room returns require more ductwork and cost more. To offset this, HVAC contractors may use pan vents through wall cavities and joist bays.

If designed correctly, a multiple-room system provides better airflow within each room, even with the door closed.

Although multiple-room return systems offer theoretical advantages, they generally don't work any better than a central system. All that additional ductwork allows for more leaks and a higher probability of system imbalance.

Fresh Air Is Important, Too

In the old days, leaky walls and windows provided ventilation by default. But as our homes have become more airtight, ventilation by design has become essential to maintain healthy indoor air quality.

Of course, the simplest ventilation involves opening a window, but thanks to heating and air-conditioning, few people ever do. ASHRAE recommends supplemental mechanical ventilation for any home tighter than .35 air changes per hour, as determined by a blower-door test. If you follow the recommendations in this book, your house will probably meet or exceed this criterion. What should you do?

You've probably already done it. Mechanical ventilation includes such commonplace appliances as bath fans and exterior venting range hoods. When you run any household fan, you mechanically ventilate your home. Few of us, however, use these commonplace ventilation systems appropriately. For example, when you shower, you should let your bath fan run for 10 to 15 minutes after bathing. Most of us shut the fan off with the light.

Because we rarely use these systems adequately, many heating and air-conditioning contractors recommend using a passive mechanical ventilation system. These can involve heat-recovery ventilators (HRVs) that warm or cool fresh air introduced into the house through a high-efficiency fan. These provide effective ventilation in summer and winter without human intervention. They cost hundreds, if not thousands, of dollars to install. But there are a few inexpensive alternatives.

> A passive way to introduce fresh air without the high cost of an HRV is to bring outside air into your return air at the furnace through a duct.

To start, you could install a high-quality, low-noise exhaust fan in a central hall and have it controlled with a switch to turn on as needed or, better yet, on a timer to cycle on a predetermined schedule. You could also run a fan off a humidistat to come on when household humidity rises beyond an acceptable limit (usually 30% to 40% relative humidity).

A passive way to introduce fresh air without the high cost of an HRV is to bring outside air into your return air at the furnace through a duct. Of course, if you just bring in outside air without a control, the air might come in hot, cold, or very damp. The solution comes with an inline damper. The damper is controlled by a dual outdoor thermostat and humidistat. Aprilaire manufactures a ventilation-control system that includes a 6-in. damper, a control knob, a transformer, and an outside temperature and humidity gauge. The damper is installed in a duct that draws outside air into your return-air plenum.

Depending on the size of your home, you select both time and cycle settings. The damper opens at predetermined times, allowing fresh air into the ventilation system. But when outdoor temperatures drop below 0°F or above 100°F, the damper closes. Likewise, it closes when relative humidity tops 60%. The system does not provide all of the advantages of a sophisticated HRV fan, but at about $200, it does an excellent job of providing affordable mechanical ventilation.

TRADE SECRETS

Excessive noise from central-air return can be minimized by using a flexible canvas connector at the return plenum.

7 Wiring and Light Fixtures

Most cities and counties follow the National Electrical Code (NEC), which limits cost-cutting options when it comes to residential electrical systems. Despite these strict regulations, there are still substantial savings to be found by planning the layout of your home from an electrician's perspective.

Start by looking at the site plan to shorten the service run, then review the arrangement of lights, plugs, and switches on the blueprint. The principles of saving money in an electrical installation parallel those for plumbing: Grouping fixtures into accessible clusters and consolidating cable runs will reduce overall costs.

Potential cost cutting is limited in electrical installations. But if you think like an electrician, you can save money by reducing the number and size of circuits, cables, outlets, switches, and fixtures.

Fine-Tuning the Site Plan

Although you may not always have a choice on where to install your meter panel, run all of your cable utilities—

In most communities, new electrical service cables are installed underground. Reducing the length of the service run saves money.

including phone, power, and television—in one trench whenever possible. Bury the sewer and water lines in another trench nearby. This will allow the installation of all utilities on one side of the house and the grouping of utility hookups in a single room. This mechanical room can be part of the garage or basement or, preferably, it can double as the laundry room.

The water service should be near the electrical panel because the electrician must connect a ground wire to the cold-water pipe within 5 ft. of the water-service entrance. When utilities are grouped in one room, the electrician can use a short length of wire to reach the water service instead of running cable all the way across the house. By placing the utility room in the laundry, along with heating and cooling equipment and the water heater, the length of expensive, heavy-gauge wire required to feed appliances can be reduced. Even the need for a separate disconnect at the furnace can be eliminated because the breaker panel is nearby.

Alternative panel locations

Sometimes there's no choice when it comes to locating the meter panel. Perhaps your water service comes through the front yard and your electrical service from the rear. The electric company may even specify an exact location for the meter, providing no flexibility at all. If this interferes with the floor plan, consider facing the breaker box toward the exterior. Some builders always use an exterior breaker box so that they can place the meter panel as close as possible to the utility service entrance, regardless of where it lands on the layout.

When it's impossible to locate the panel near the water service, provide your electrician with a concrete-encased electrode in lieu of a ground using cold-water lines. Leave a few inches of rebar poking out of the stem wall within easy reach of the panel location so the electrician can clamp the ground wire to it.

For the secondary ground wire, also required by code, you can bury a 20-ft. length of no. 2 bare copper wire around the perimeter of the house. Just drop the wire in the area dug

An exterior panel provides ultimate flexibility, allowing the meter to be placed as close as possible to the service entrance without disrupting the interior floor plan.

TRADE SECRETS

Code requires all household smoke detectors to be interconnected by a third wire, but the code exempts small single-story homes in which occupants can hear any of the alarms when they sound.

A LOW-COST ALTERNATIVE FOR GROUNDING

A low-cost grounding electrode can be fashioned from a steel reinforcement bar ½ in. or larger in diameter and at least 20 ft. long. It should be encased in at least 2 in. of concrete and located within the bottom of a concrete footing. Called an Ufer grounding rod after its inventor, the rebar is set in place before concrete is poured for the footing.

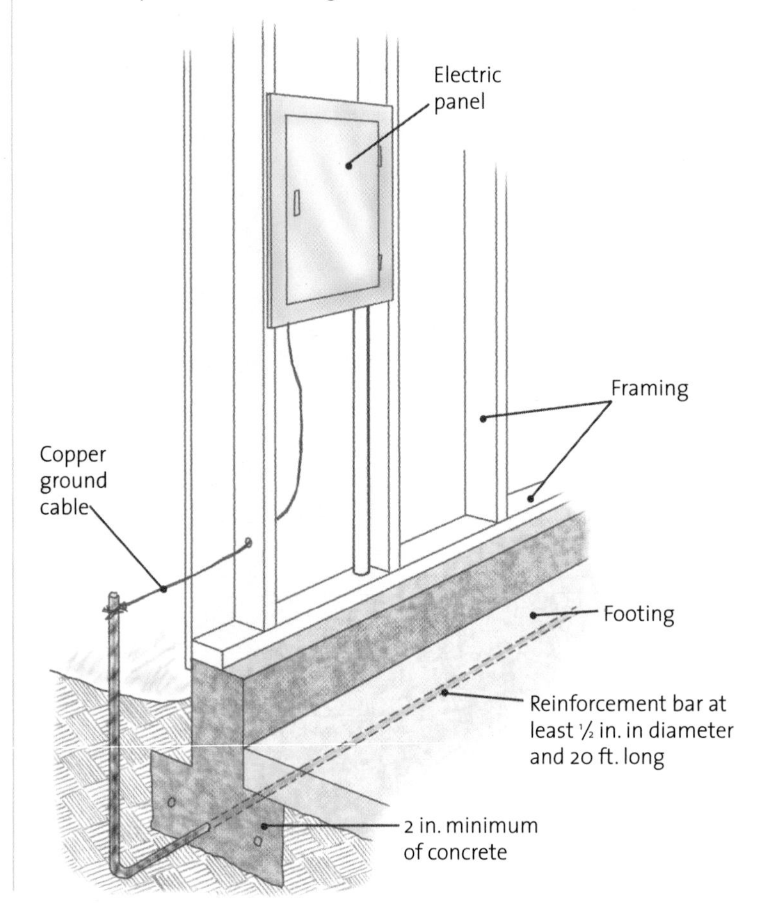

out for the basement before it's backfilled, making sure it's at least 30 in. deep. Although this does not save on expensive copper wire, it's a lot less labor intensive than drilling a similar length of cable through an array of studs to reach the water service.

Devising a Smart Floor Plan

One of the best ways to reduce electrical costs is to reduce the number of outlets and switches. Since the electrician will install a receptacle about every 12 ft., shortening or even eliminating walls can purge outlets at a savings rate of about $55 each. (Your eraser can be a very lucrative drafting tool.) Check plans for arbitrary outlet locations and eliminate ones that are unnecessary. For example, there's no need to install an outlet in a hallway less than 10 ft. long. Whenever possible, relocate closet doors to avoid 24-in. lengths of wall, which require an outlet by code. Reduce this length by an inch and you can avoid an outlet.

TRADE SECRETS

In certain rooms, such as the living room or bedrooms, a switched outlet for a floor lamp or table lamp can replace cable runs to the ceiling for an overhead fixture. That reduces wiring and fixture costs simultaneously.

Choosing Breakers and Boxes

Breakers come in two price ranges, residential and commercial. There's no reason to choose the more expensive variety because every circuit breaker must meet the same Underwriter's Laboratories (UL) standards.

Although all breakers and panels must conform to the same UL standards, avoid the lure of inexpensive circuit boxes with equal capacity but less room for individual circuit breakers. For example, a 200-amp breaker panel with room for only 20 full-sized breakers might seem like a bargain until you run out of capacity. Sure, you can use double-pole breakers but these cost more than two single breakers and add substantial expense to the total cost of the installation.

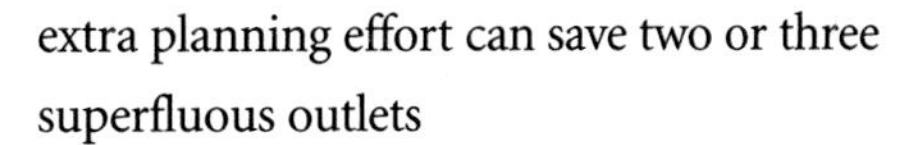

By placing receptacles off center on a wall, you can take advantage of access to an outlet from two directions to minimize the total number of receptacles in a room. Ideally, each living area will require only three or four outlets; if you have more, double-check the design. Prescriptive code calls for an outlet every 12 ft., but outlets can be located at points of probable use if their placement is presented in a well-thought-out plan to the code-enforcement official for review. If you plan carefully, you might be able to delete an outlet even when the 12-ft. rule would seem to require it. The National Electrical Code allows some wiggle room if plans meet the intent of the code, which stipulates receptacles in all locations of likely use. In many cases, this extra planning effort can save two or three superfluous outlets

Since the electrician will measure outlet locations starting at 6 ft. from the operable side of a door, make sure a stud is located there, to which the box can be nailed. Otherwise, the electrician will have to use the next closest stud, possibly adding an unnecessary outlet to the room.

> One of the best ways to reduce electrical costs is to reduce the number of outlets and switches.

Careful planning in the kitchen also can help. Any countertop 12 in. or longer requires an outlet. When you have an isolated countertop, such as a short counter next to the refrigerator, make sure it measures only 11 in. Likewise, keep kitchen islands and peninsulas under 24 in. in length to avoid extra plugs. Adding extra blocking in the walls is another way to reduce the number of kitchen receptacles.

ADD A KITCHEN LEDGER

To help the electrician avoid extra kitchen receptacles, install a ledger of 2x4 blocks at 43 in. along the back of every counter. This allows the electrician to locate receptacles exactly where required, not at the closest stud, saving one or two outlets per kitchen.

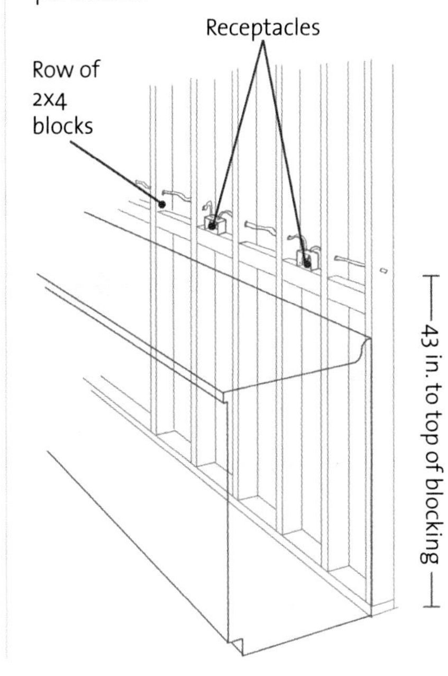

ONE RECEPTACLE EQUALS $55

The left panel shows a standard bedroom layout with four receptacles. The right panel shows the same room with only three receptacles when configured to optimize the electrical installation.

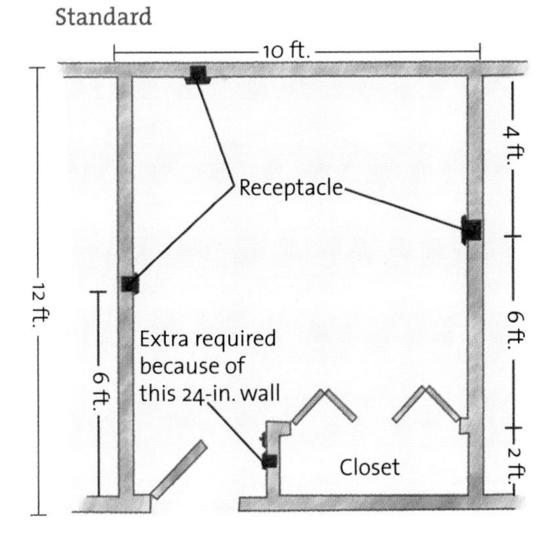

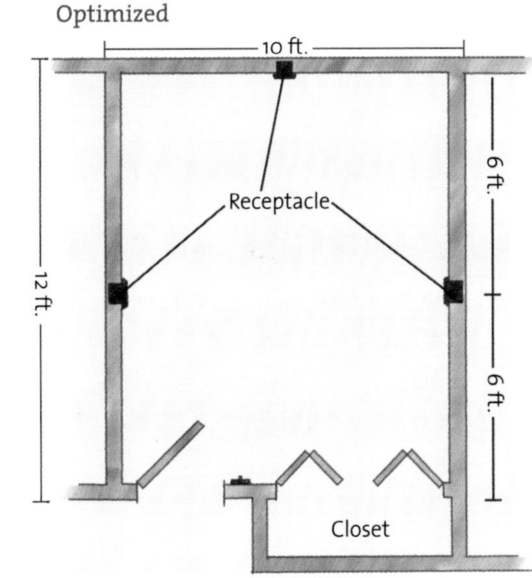

Money-Saving "Green" Checklist for Electrical

- Install compact fluorescent lightbulbs.
- Install insulation-compatible recessed lighting.
- Install ceiling fans.

Wiring more than one light to a single control is one way of reducing the number of switches, thus saving money. But make sure the arrangement isn't inconvenient to use.

Reducing Lights and Switches

Within reason, several lights can operate from a single control, but it's wise to make sure the layout won't be inconvenient to use. For example, a bath fan and light require only one toggle, but many people prefer to control the fan separately. If you're going to use a fan/light combination without a separate switched light, use one with a very quiet motor.

> Within reason, several lights can operate from a single control.

On the other hand, multiple switching options are rarely needed in small houses with short hallways and modest rooms. A single switch set midway in a short hall makes an expensive three-way switch unnecessary. There's no need to install two switches for a ceiling fan. Ceiling-fan motors can operate from a pull chain or a remote control, with only one wall toggle supplying power to the entire fixture instead of two separate switches for the fan and the light.

Pull-chain fixtures also perform adequately in attics, storage spaces, and certain mechanical rooms, eliminating the need for switches entirely. In jurisdictions where codes still allow it, a porcelain light with a plug socket can perform double duty by providing an extra convenience outlet. Wardrobe fixtures also can be omitted since only particularly large walk-in closets actually require more light than what travels through the open door.

If you use a decorative door knocker, you can avoid wiring for an electric doorbell. This may seem trivial, but consider that the doorbell requires a transformer outlet, low-voltage wiring, and another outlet for the chime.

Although electricians often install several lights in utility rooms, such as garages and basements, codes only require one. Likewise, only one outlet in each utility area and one outlet for the exterior are required. For utility lighting, use a simple porcelain lamp holder, which works as well as any other fixture and costs less than $2.

TRADE SECRETS

If you want to save energy, use standard incandescent lamp holders with screw-in fluorescent bulbs instead of supplying expensive fluorescent tube fixtures.

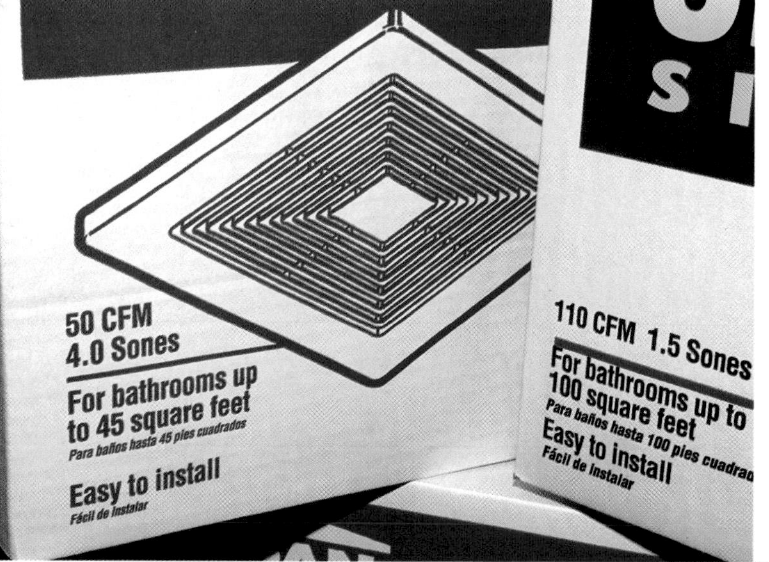

Money saved by buying an inexpensive, noisy fan isn't worth the aggravation. Noise is measured in "sones." The higher the number, the louder the fan. Look for a low-sone unit that ducts to the outside.

THREE-WIRE CABLE DOES DOUBLE DUTY

Two circuits on this second-floor plan can be fed by a single three-wire line: The three-wire cable splits into two circuits of two-wire cable at an outlet box, saving the cost of running two individual circuits all the way from the breaker box to the second floor.

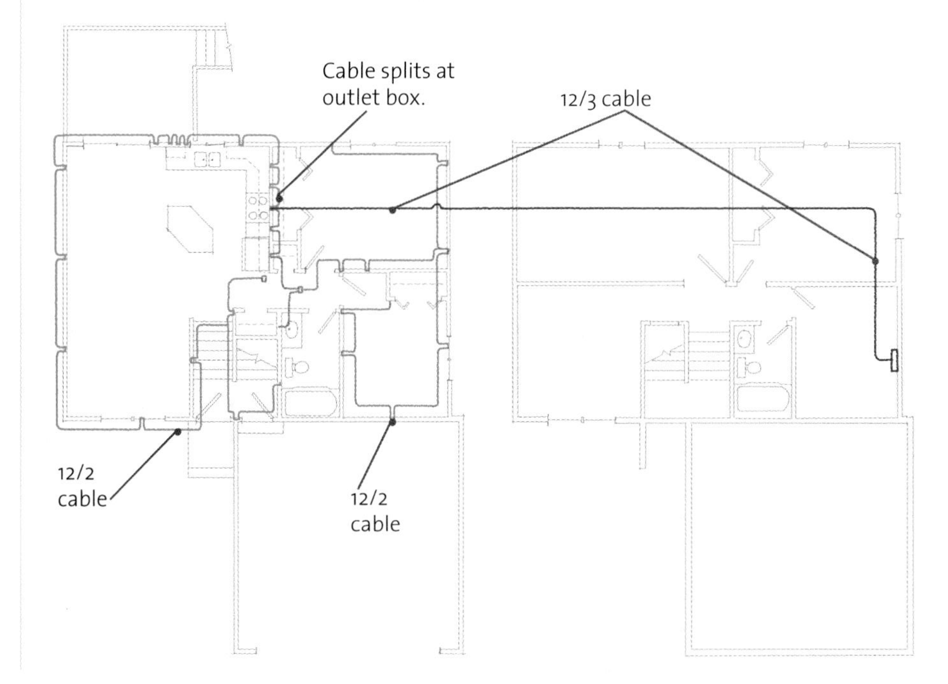

You can save money right-sizing your electrical distribution system. Many smaller homes can be served with a 100- or 150-amp panel.

Choose Wire Gauge Carefully

Many homes are wired with 12-ga. wire and 20-amp circuit breakers for general wiring, while codes allow 14-ga. wire and 15-amp circuits for most outlets and fixtures. Unless 12-ga. wire is specifically required, the lighter 14-ga. wire will be less expensive to use. In the same vein, 200-amp service panels have become standard. But unless it has electric heating devices and heavy appliances, a smaller home can be adequately served with a 100- or 150-amp load center instead.

Although wires and circuits should never be oversized, sometimes it's less expensive to run 20 lamp holders and receptacles on a single 20-amp circuit than it is to wire 15 devices off a 15-amp breaker. This makes especially good sense when you're wiring on a second floor or in any area of the house that's far from the main panel. Unless you maximize the number of outlets on each circuit, oversizing simply costs more without adding tangible benefits.

> Although electricians often install several lights in utility rooms, such as garages and basements, codes only require one.

Electricians routinely install extra branch circuits to simplify wiring, but a multiwire (three-wire) cable allows you to maximize the number of devices on a single circuit while reducing wiring. Each cable with three conductors instead of the more common two (14/3 or 12/3 cable versus 14/2 or 12/2) can feed two circuits of 15 to 20 devices.

BUILDER'S CORNER

Take Care with Wall and Ceiling Boxes

Before the drywaller installers arrive, mark every wall- and ceiling-box location on the floor with a lumber crayon. After completing the drywall, have your laborers, instead of your electricians, clean drywall compound off the boxes and check that your installers haven't buried any outlets. If they have, ask the installers to find and expose these before the electrician arrives. Make sure that drywall installers don't cut wires while trimming around electrical boxes with a router. Routers also love to nip the screw tabs off plastic plaster rings, forcing your electrician to break the wall and replace the ring.

TRADE SECRETS

Buy 14/3 and 12/3 cable in 1,000-ft. spools. Rolls of 250 ft. or less leave short (under 15 ft.) lengths of wire at the end of each roll. Although a 1,000-ft. spool of wire doesn't cost less per foot, you can save up to 30 ft. of wasted wire scraps per roll.

Using three-wire cable (at right) instead of more common two-wire sheathed cable allows two circuits to be powered by a single line from the service panel.

Aluminum represents a safe and economical alternative when used to feed single-purpose high-amperage circuits, such as 240 volt air-conditioning equipment and electric ranges.

Wiring receptacles in series to a single GFCI is less expensive and just as safe as providing ground-fault devices at every outlet.

Ground-Fault Circuit Interrupters

You can loop-wire ground-fault circuit interrupters (GFCIs) so several outlets are wired in series from a single GFCI device. Likewise, loop off a GFCI outlet for each kitchen circuit. Although it might seem that running GFCI-protected circuits from a GFCI breaker in the box instead of looping individual outlets should cost less, it doesn't. These specialty breakers are very expensive.

Although code requires separate circuits for bathroom receptacles and lights, where a bathroom has only one outlet, the NEC permits a single 15-amp circuit (in some cities 20-amp circuits are required) to feed the outlet along with vanity lights and a fan.

> Aluminum represents a safe and economical alternative when it's used to feed single-purpose, high-amperage circuits

Consider Aluminum Wire

Electricians no longer use aluminum for standard household wiring in the United States, due to a rash of electrical fires in the 1970s that were caused by faulty connections. But it's still being installed in Canada with special copper/aluminum devices that compensate for the thermal shrinkage of aluminum. Aluminum represents a safe, economical alternative when it's used to feed single-purpose, high-amperage circuits, such as 240-volt lines for air-conditioning equipment and electric ranges.

It can also be used to feed subpanels, such as those used for electrical distribution in a detached outbuilding or duplex apartment. Using sheathed, nonmetallic aluminum cable to connect the master panel with higher-amperage appliances and subpanels saves labor and material costs over nonmetallic copper cable or conduit and pull wire. Aluminum is less expensive than copper, and using sheathed cable is less expensive than pulling wires through a conduit.

Avoid unnecessary drilling

Where possible, rope your cable over corners and through the attic instead of through gangs of studs. This uses extra wire, but it's faster, saves drilling through corners, and requires fewer protection plates. If your electrician can offer a price reduction to offset any additional framing costs, consider framing with predrilled studs or an engineered joist system with knock-outs for wiring. In utility areas, such as the laundry room, garage, and closets, plan raceways so the electrician can run cables within a framed chase instead of drilling through joists and studs.

I-joists usually come with scored punch-outs for wiring, which saves the electrician time drilling joists. Make sure you get credit for the time your electrician saves over conventional framing, helping offset the higher cost of engineered joists.

Drilling all those holes is a lot of work. Instead of having your electrician auger through joists, provide a framed raceway.

A SIMPLE RACEWAY FOR CABLE

Instead of forcing your electrician to auger through miles of joists, provide a framed and drywalled chase to encase cables.

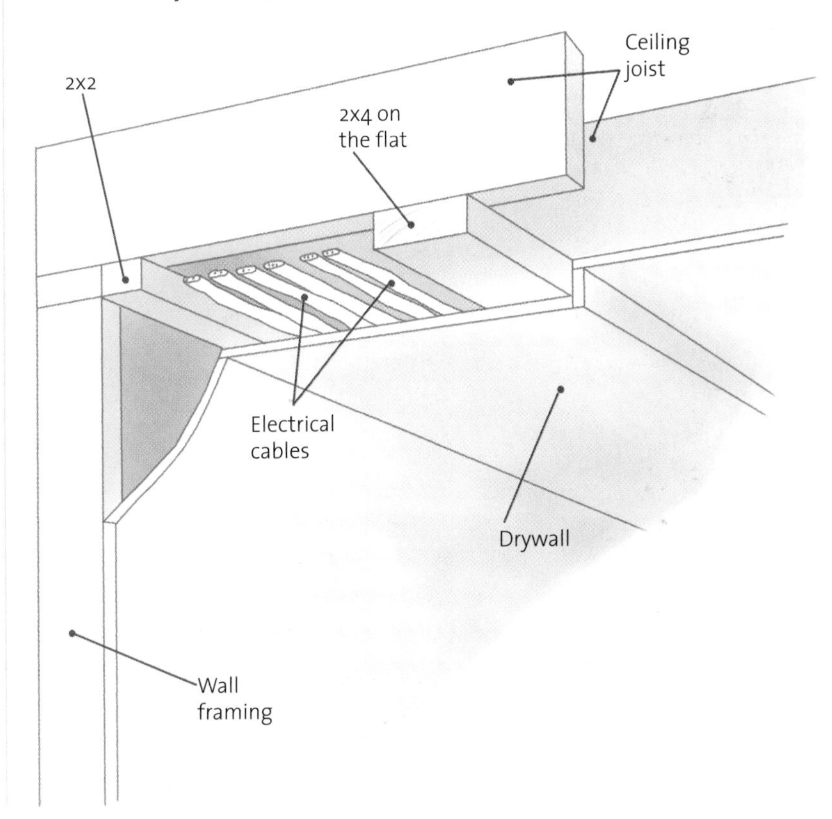

Recessed Fixtures Save Money

In many instances, a recessed fixture can save money in labor. Although a recessed light costs about $10 more than an economy light fixture, the recessed light installation is essentially finished at electrical rough-in. Recessed lights only require a bulb and trim ring to complete. These fixtures don't require a junction box, so there's an additional savings of $1.50 in materials and a few minutes of work. Recessed lights save about 15 minutes of work each, which, at an electrician's wage, represents a net savings of almost $10 per light.

In many instances, a recessed fixture can save money in labor.

Try to avoid expensive fixtures, such as the four-tube fluorescent lights that are often installed on kitchen ceilings. Although code requires that the first switch you reach as you enter the kitchen turns on a fluorescent fixture, you can connect this switch to an under-cabinet fluorescent light or, even better, an economical recessed light over the sink fitted with a fluorescent bulb. Then you can use a less expensive, more stylish incandescent lamp for your principal overhead lighting.

A combined fan and fluorescent light for the bathroom can actually cost more than separate fan and light fixtures. But when you factor in the extra switch, wire, junction box, and labor required to wire two separate fixtures, the combination device actually turns out to be less expensive.

To save money, use a simple, old-fashioned toggle switch and standard plugs. But before deciding on the color, check with your supply house to see if they have overstock. As fashion trends shift from one color to another, you may be able to find white, ivory, or almond switches, plugs, and face plates on sale.

TRADE SECRETS

Use a sink base cabinet with no back so the electrician can use a simple rough-in box without an extension, and the back of the cabinet does not have to be marked and cut.

Installing a recessed light saves about 15 minutes of work, a savings of about $10 at an electrician's wage.

BUILDER'S CORNER

Avoid Confusion Over Light Fixtures

NuHome in Houston installs only two types of light fixtures throughout its homes. One fixture has two lamp holders, the other has one. Both consist of a polished brass base with a white shade. Depending on the location, the electrician knows which fixture to install: Bedrooms and kitchens get the two-bulb version, bathrooms and hallways the single bulb. By buying in bulk, the fixtures cost even less and the electrician never gets confused. As they can afford to, buyers upgrade their fixtures.

Reducing light fixtures to one- or two-bulb models reduces confusion for the electrician while saving money. NuHome of Houston uses a simple, two-bulb incandescent fixture in the kitchens of the houses it builds.

TRADE SECRETS

When you wire a disconnect for a compressor, use a combination disconnect and GFCI-protected utility outlet so you don't have to run two separate lines.

TRADE SECRETS

In many jurisdictions, you can use a well-supported pancake or metal ceiling box in lieu of an approved fan outlet. Check with your local inspector.

Oakwood Homes of Denver uses a two-bulb incandescent ceiling fixture as an inexpensive vanity light.

8 Insulation

Spending a little extra money on insulation and air sealing can provide substantial returns in quality comfort and reduced operating expenses. At the same time, it's possible to build an energy-efficient house affordably. If you have followed the advanced lumber-sparing techniques described in Chapter 4, you're well on your way to a better insulated home.

With wood framing making up about 30% of the exterior surface of a typical house, a substantial reduction in lumber can improve energy performance dramatically. Just consider that the least expensive fiberglass batt insulation has an R-value of 3 to 4 per inch, while wood has an R-value of about 1 per inch, and you'll understand why less wood per square foot plus more insulation equals a more energy-efficient home.

Fiberglass batt insulation is relatively inexpensive, but it will perform well when installed carefully. It should be cut to fit snugly without reducing its loft.

Spend Money to Save Money

Simple air-sealing practices, such as installing air-barrier tape, chinking window and door frames with fiberglass insulation, caulking between double studs, applying foundation sealer, and foaming wire and pipe penetrations at the plates, can add $100 to $200 to the cost of a house. But it's money well spent.

These measures do a great deal to improve thermal performance even when using fiberglass batts. Air sealing, in fact, is more effective than using insulation with a higher R-value or choosing a more expensive type of insulation, such as the fiberglass Blown-in-Blanket System or blown-in cellulose.

In tests conducted by the NAHB Research Center, a well-sealed home insulated with fiberglass batts performed as well as one insulated with blown-in fiberglass and only negligibly worse than one insulated with cellulose. The air-sealing techniques described here have a payback period averaging one or two years, where the cost of spending more for higher R-value insulation can take between 20 and 40 years to recover.

TRADE SECRETS

In comparative tests, attic fans don't work any better at controlling moisture than passive systems. If you're concerned about moisture levels in the attic, install an impervious air barrier in the ceiling beneath the attic instead of a more elaborate ventilation system.

As consumers have become informed about the importance of insulation, manufacturers have introduced new, cleverly designed proprietary products to improve the efficiency of the building envelope—for a price. But many of the best methods also happen to be the least expensive, including the use of expanding foam sealants, fiberglass batt insulation, and caulk.

Making the Building Tight

R-value alone does not guarantee comfort or performance. A carefully sealed wall insulated with R-13 fiberglass batts can perform better than an R-19 wall where no special measures have been taken to control air leakage. Higher insulation levels can actually increase condensation within walls, as moist household air is forced into wall cavities whenever outside surface temperatures drop below the temperature of inside surfaces. Without proper air-sealing, higher insulation values exacerbate this problem by creating a greater temperature difference, inducing even more air movement into the wall cavity. Making the exterior shell airtight—instead of paying for increased R-values—provides the best and least expensive means for more comfort and exceptional energy performance.

> Many of the best methods also happen to be the least expensive

There are two basic approaches to controlling air leakage: a polyethylene vapor barrier applied to the interior face of the walls and a vapor-permeable polyolefin paper, such as DuPont's Tyvek™, on the exterior. Fiberglass batts with a facing of kraft paper have fallen out of favor. But truth be told, kraft paper is actually a better indoor option than polyethylene. In a step back to the future, building scientists have starting steering homebuilders away from the plastic-bag approach because poly wrap makes moisture problems worse, and it fails notoriously in pressure tests.

It's difficult to install an air barrier inside the house, where electrical outlets, plumbing penetrations, stringers, intersecting walls, and even bathtubs all can get in the way. Throughout most of the United States, it's

Controlling air leaks by filling gaps around windows and foaming or caulking electrical and plumbing penetrations is a big part of ensuring overall thermal efficiency.

In most parts of the country, polyethylene plastic installed as a vapor retarder is more trouble than it's worth. The thin material is easily damaged during construction, and it can trap moisture inside wall cavities.

Fiberglass batts faced with kraft paper provide an excellent moisture barrier that protects the wall cavity from infiltration and allows the materials within the wall to breathe.

appropriate to forgo the interior poly wrap and concentrate air-sealing efforts on the exterior. Inside, kraft paper provides an intelligent second-line moisture barrier, which, unlike polyethylene, does not seal humidity into the wall cavities and create the potential for mold. Kraft paper breathes, and it's less expensive.

Whatever brand of exterior barrier you choose, it's important to install it correctly.

An exterior barrier installs easily, especially around window and door openings where most infiltration occurs. Some exterior barriers, such as Tyvek Stucco Wrap™, have a corrugated surface that provides a built-in drainage plane that channels water and moisture to the outside. You can use this stucco wrap under any siding.

Whatever brand of exterior barrier you choose, it's important to install it correctly. Fortunately, window flashing and seam tape have made the operation easy. It's not necessary to use expensive, proprietary products, except in those instances where the flexibility they provide around curves and architectural details makes for quick work in otherwise difficult flashing situations. Otherwise, traditional window flashing and building paper work just as well.

Although experts don't recommend polyethylene on walls, it is effective on the ceiling below the attic if properly sealed around light boxes and bath fans. Canadian tests show that properly installed polyethylene or the Airtight Drywall Approach (see p. 152) can reduce attic humidity enough so attic ventilation is unnecessary. This is a useful option for homes without eave and ridge vents, or whenever the efficiency of the attic ventilation system becomes questionable. In fact, if the attic junction is nearly airtight, standard approaches to attic ventilation—including eave and roof vents—actually result in higher moisture levels by drawing humid outside air into the roof framing.

Plug the Small Gaps, Too

If you have any doubt that electrical outlets leak enormous amounts of cold air into your houses, ask an energy consultant to run a blower-door test. When the house is depressurized, put your hand up to a switch plate or receptacle and prepare for a startling realization: Air runs down wires and pipes and leaks into your house just

Installing a moisture barrier in the ceiling immediately below the attic can make attic ventilation unnecessary and lead to savings in time and materials.

AIR-SEALING EXTERIOR WALLS

House wrap doesn't stop the movement of water vapor, but it does prevent wind from washing through wall cavities and reducing the efficiency of insulation. It should be applied according to the manufacturer's recommendations.

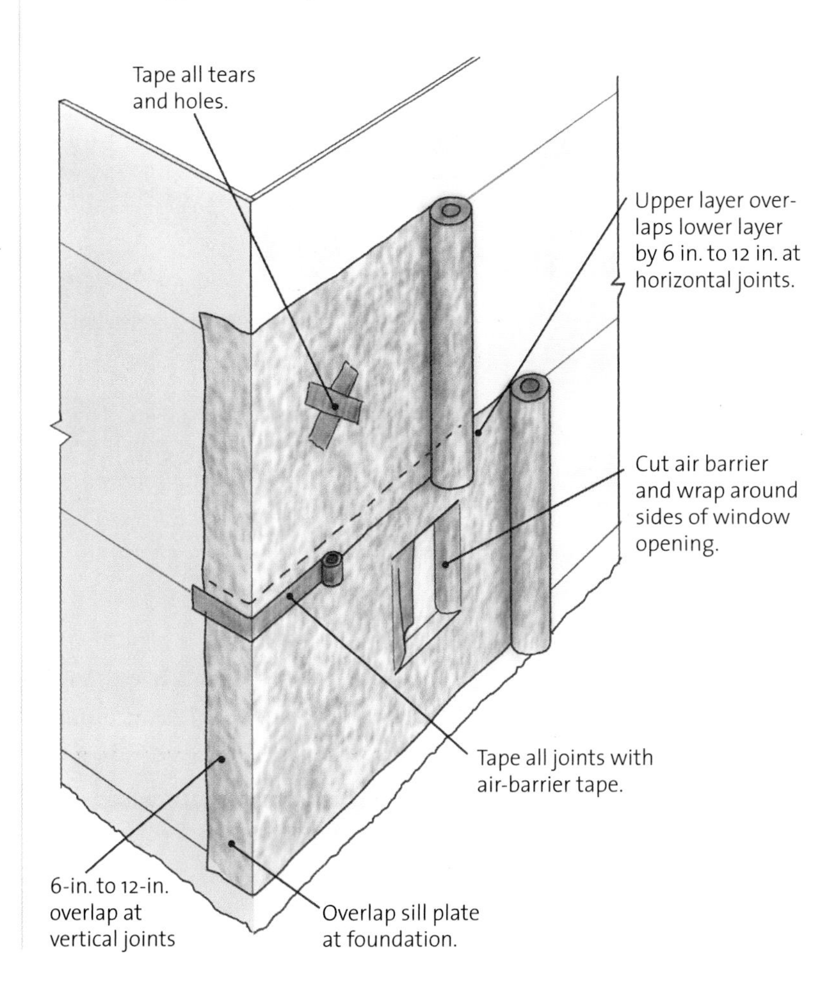

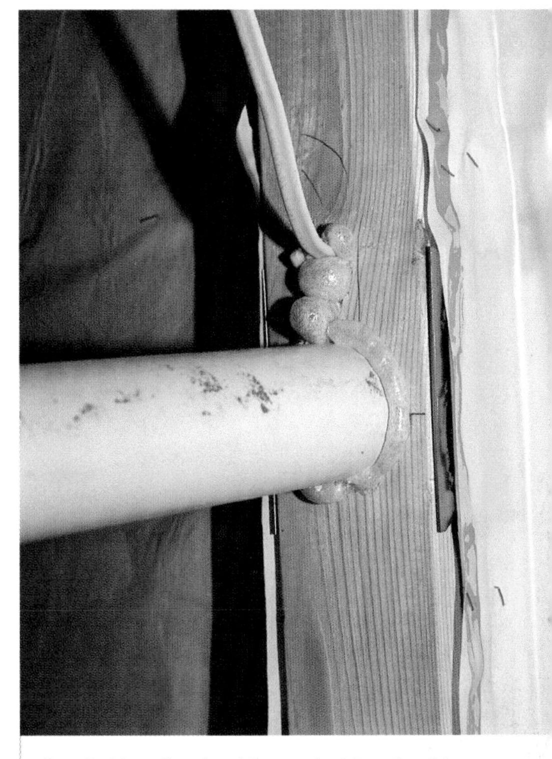

Penetrations for plumbing and wiring should be sealed with expanding polyurethane foam or caulk to reduce air infiltration.

like water. To keep air from leaking through wall and ceiling penetrations, use caulking and expanding foam sealants.

You may be persuaded that blown insulation can replace proper sealing, but in head-to-head tests conducted by the NAHB Research Center, a properly sealed home insulated with fiberglass batts had less infiltration than blown-in fiberglass, spray cellulose, or foam-in-place insulation alone. Besides, sealing penetrations provides one of the least expensive means to improve your building practices. If you do nothing more than foam gaps where pipes and wires go through your top plates to the attic and caulk around plugs, switches, and light boxes at exterior walls, you can improve your home's energy performance by 30%. That may be enough to reduce the size of the air-conditioning compressor by a half-ton, and the furnace by 50,000 Btu (see Chapter 6).

> Sealing penetrations provides one of the least expensive means to improve your building practices.

TRADE SECRETS

Two air-sealing efforts that result in a consistent short-term payback include sealing around windows with air-barrier tape and installing a basic interior air-sealing package. The basic package includes caulk at top plates and double studs, foam at electrical and plumbing penetrations, and the use of a sill-sealer along the foundation.

After installing drywall, caulk the joint between the drywall and any electrical boxes in the ceiling to prevent air infiltration.

Special electrical rough-in boxes can help provide a good air seal around switch and outlet boxes, a common source of air leaks.

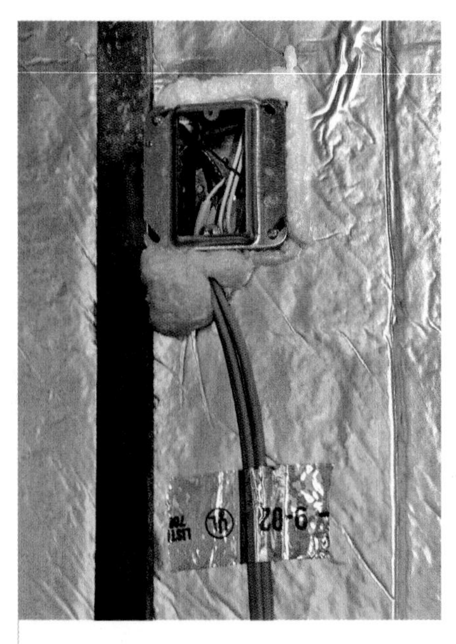

Use an expanding-foam sealant to block potential air leaks at all outlets, even on lower floors. Cold air travels through floor cavities and down wires just like water.

Sealing up holes in electrical switch and outlet boxes can help keep out cold air. The practice is inexpensive but effective.

Several specialty products are available that can provide a positive seal around electrical outlets, such as plastic electrical rough-in boxes with a flange for caulking. Much the same benefit is possible simply by foaming wires where they come into ceiling junction boxes and caulking ceiling and exterior wall outlets tight to the finished drywall.

The challenge is in the details

Besides windows and doors, the leakiest areas of a home are the perimeter of bathtubs and chimneys. This is another area where a builder can make a huge difference in the insulation of a home by batting and sealing the wall behind a tub before installing the tub. Foaming around the fireplace frame and insulating around the fireplace chase keep drafty, cold pockets from developing.

Double studs and corners also allow cold air into the house, but a bead of caulk keeps it out. Many insulation experts also recommend caulking along the sill and double top plate on exterior walls.

Perhaps the most difficult framing assembly to seal and insulate is the rim joist. Insulation often falls out of the space between joists during construction. One solution is to recess the rim joists and laminate them with high-density foam board. Another is to conduct an insulation inspection right before drywall is hung to verify that all the perimeter insulation batts are firmly in place and installed correctly. If you do your own caulking and weather sealing, you can caulk between the sill plate, rim joist, and floor sheathing and then install the perimeter insulation right before drywall. (This is not a bad project for an owner to do when having a house custom built.) It's an easy job, but just as easily overlooked. Some builders opt to have the interior face of the rim joists foamed with low-density polyurethane, which provides an excellent solution, albeit an expensive one.

Spray urethane foam is an easy solution but a relatively expensive material for sealing the rim joist. There are several low-cost approaches that also can be effective for this common trouble spot.

Gaps around window and door frames should be insulated carefully, but be careful not to overcompress the insulation. Fiberglass insulation requires loft to provide any benefit, and too much insulation may bow the frame.

SEALING THE RIM JOIST

Air leaks and heat loss are common problems at the rim joist. There are several ways of tackling the problem with a variety of insulating materials. The key is careful installation.

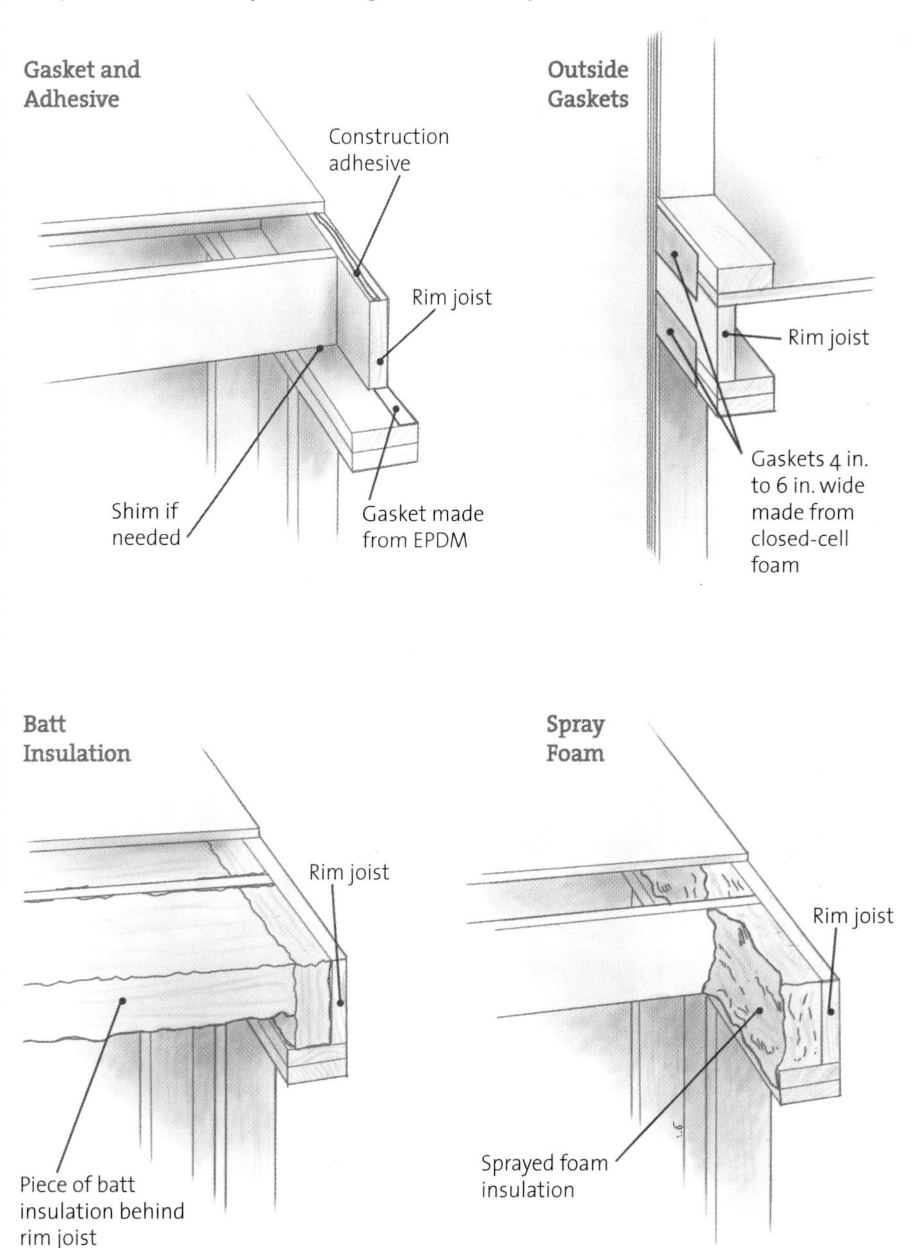

TRADE SECRETS

Builders should teach their employees to install the sealing package. They will deliver a more consistent and careful job than the insulation contractor, and generally for a better price.

The Airtight Drywall Approach

Drywall accounts for about 70% of the interior air seal, but it can be more efficient than that. In Canada, building scientists devised a drywall-based air-sealing system called the Airtight Drywall Approach (ADA). It involves the use of gaskets and seals along the exterior edges and penetrations of drywall surfaces to provide an air barrier. Only a few builders in the United States have adopted this system, but it's worth learning about because it's one of the most effective, easiest, and least expensive means of sealing a home.

> Drywall accounts for about 70% of the interior air seal, but it can be more efficient than that.

Like polyethylene, ADA is a moisture barrier at the interior surface, which is appropriate in colder climates. But unlike poly, drywall does not fail under wind pressure. It's not easily torn, and you can seal it tightly around electrical outlets and plumbing penetrations with caulk.

This approach uses drywall, industrial gaskets, and caulk to create a continuous air retarder. The first step is to seal places where the foundation, sill plate, floor joists, header, and subfloors meet, just as you would with any good sealing system. Then ADA adds a pressure-adhesive, closed-cell foam or EPDM gasket between the framing and drywall on exterior walls and ceilings. The end stud and top plates of interior partitions get the same. Many builders use special electrical boxes with flanges and gasketing or caulk to provide an unbroken seal.

Money-Saving "Green" Checklist for Insulation

- Upgrade sealing package with advanced infiltration-reduction methods.
- Use recycled cellulose-based blown insulation in attics.
- Use recycled formaldehyde-free fiberglass insulation in walls.

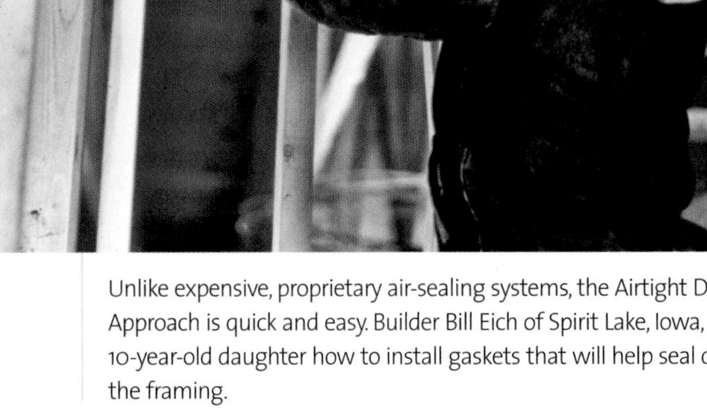

Unlike expensive, proprietary air-sealing systems, the Airtight Drywall Approach is quick and easy. Builder Bill Eich of Spirit Lake, Iowa, taught his 10-year-old daughter how to install gaskets that will help seal drywall to the framing.

AIRTIGHT DRYWALL APPROACH

The ADA uses drywall and simple, self-adhesive gasketing material to create a tight, continuous air barrier more effective than polyethylene.

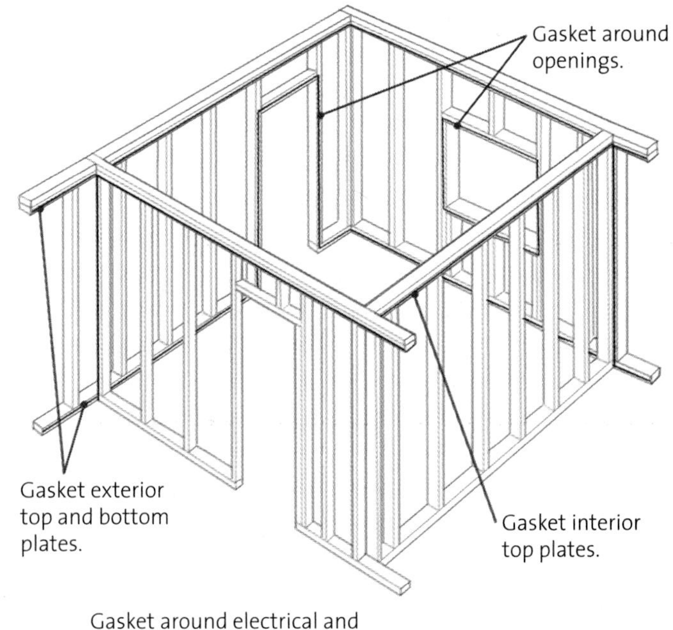

Fiberglass batt insulation comes with or without kraft-paper facing. It's also available with a foil facing, which is less permeable to moisture than the paper.

Choosing the Right Insulation

Insulation products vary substantially in cost, with fiberglass batts the least expensive and foam-in-place polyurethane the most expensive. High-quality results are possible with fiberglass batts as long as the installer pays close attention to the details, which include loft and fit.

Batts need room to expand

Fiberglass by itself has very little insulating value; it's the air trapped between fibers that provides its insulating qualities. When fiberglass obtains its full loft, batts provide the R-value indicated by batt thickness. As such, R-30 requires 10 in. of loft, R-19 requires 5½ in., and R-15 requires 3½ in. of space in which to expand. When compressed, batts can lose more than half of their insulation value.

Any gaps between batts and framing reduce thermal performance, making a good, tight fit extremely important. Batts should be measured, cut, and fitted with all the attention

TRADE SECRETS

Johns Manville has a perforated wall insulation product that literally tears along the dotted line to provide greater flexibility and ease when trimming insulation to fit in nonstandard wall cavities.

ADA: Big Results for Little Money

Although most builders use the Airtight Drywall Approach as an energy-saving technique, Joe Lstiburek, a Canadian building scientist, originally devised it in 1980 to keep airborne moisture from penetrating wall cavities. Test houses where this approach was used required 23% to 32% less heating than conventionally built homes of similar specifications.

The highly affordable approach involves many steps that collectively are very effective at controlling air leaks.

FLOORS

> Seal the rim joist to minimize air currents.

> Seal seams between subflooring with construction adhesive.

EXTERIOR WALLS

> Install gasket or caulk at the top and bottom plates of exterior walls.

> Seal penetrations through the top and bottom plates for plumbing, wiring, and ducts.

> Provide a vapor barrier with kraft-faced insulation batts.

> Use drywall joint compound or caulk to seal the seams between drywall and electrical boxes.

PARTITION WALLS

> Seal drywall at either top or bottom plate with a gasket or caulk.

> Seal penetrations through the top and bottom plates for plumbing, wiring, and ducts.

CEILINGS

> Seal all penetrations in the ceiling for wiring, plumbing, ducts, and attic access openings.

> Use whole sheets of drywall to minimize joints.

> Seal the junction between ceiling and walls with drywall compound and tape.

DOORS AND WINDOWS

> Seal drywall edges to either framing or jambs with gasket or caulk.

> Fill the space between the jambs and the framing with backer rod and caulk or insulation pressed in with a putty knife.

> Caulk casing to drywall.

Spray urethane foam insulation expands on contact to fill wall and ceiling cavities quickly, and it can provide an effective air block. But it is the most expensive of all insulation options.

that trim carpentry gets. Otherwise, air convection combined with air leakage will undermine the efforts. The inherent difficulty of installing batt insulation correctly has prompted many builders to opt for blown-in insulation products even though they are more expensive. Installed correctly, these products stop most air leaks as well as provide thermal insulation. But they don't do as good a job as sealing every crevice by hand.

Blown insulation

Cellulose is the least expensive of the blown insulation alternatives, followed by the fiberglass Blown-in-Blanket System (BIBS). Foam-in-place polyurethane systems represent the high end in cost, but not the most consistency in quality. In head-to-head tests, it was the fiberglass blown-in-blanket system that proved most reliable. Perhaps this is because it doesn't require tooling after installation. A mesh holds the insulation in place until drywall is installed.

With foam-in-place insulation, workers must either trim excess foam where it expands beyond the wall cavity with a handsaw or shoot a level depth by eye, which often yields a sloppy, uneven installation. Cellulose requires tooling, too, and can flake off and leave voids as drywall nailing vibrates the walls. Cellulose also

Any uninsulated gap, no matter how small, allows a surprising amount of cold air inside a home, making blown-in cellulose popular with many builders.

Like spray-in foam, damp-spray cellulose must be trimmed back to the wall studs with a special tool. If covered with drywall before it has thoroughly dried, the insulation can leave moisture in wall cavities.

introduces moisture into the frame, which, if covered with drywall too soon, can create long-term humidity problems. Although blown products generally score higher in insulation inspections than fiberglass batts, voids behind electrical boxes and inconsistent density are common problems to watch for, especially with foam products.

Put cellulose in the attic

In the attic, blown insulation and especially loose-fill cellulose surpass fiberglass batts in both price and performance. Because of its density, loose-fill cellulose achieves an R-value of 30 with about 8 in., compared with 13 in. for fiberglass. This means that cellulose can provide higher insulation values near the eaves and other low-clearance areas. Because it remains exposed to the air, the higher moisture content found in cellulose doesn't create long-term

Performance without the Problems

Blown foam insulation has always had environmental drawbacks. The chlorofluorocarbons (CFCs) that manufacturers used as a blowing agent depletes the ozone layer in the atmosphere. Even the new generation of hydrochlorofluorocarbons (HCFCs) is not entirely benign. But an emerging group of soy-based polyurethane products promises an environmentally safe solution.

Soy-based foam insulation produces no HCFCs or CFCs, and it does not contain formaldehyde or any other volatile organic compounds. The soy-based products cost less than other foam insulation systems, and they do not rely on foreign energy producers. Industry literature claims that soy-based products can compete in price with fiberglass batt insulation—the least expensive kind of insulation—but I have not found an installer who confirms this. Soy products may cost less, but any type of foaming takes almost two-and-a-half times longer to install than batts. And time is still money.

In the Blown-in-Blanket System, netting applied to studs contains loose-fill fiberglass.

TRADE SECRETS

Many builders pay their insulation contractors by the square foot to ensure a low price, and they get it. You might not save money by buying your own batt insulation and installing it—or by hiring the best subcontractor available—but you will be assured of a better job.

Loose-fill cellulose insulation costs less than fiberglass batts. In the attic, cellulose seals around pipes tightly and provides higher R-values per inch than fiberglass.

An economical approach to furring concrete basement walls: Drywaller installs insulation board held in place by metal Z-bar. The metal not only holds the insulation in place, but also provides a nailing flange for future drywall.

Soy-oil-based polyurethane foam insulation not only represents an environmentally responsible alternative, it also costs less than petroleum-based foam products.

humidity concerns. It also settles around pipes and wires to form a tighter fit. But don't be fooled into thinking you don't need to foam plate penetrations—cellulose provides an excellent R-value, not air sealing.

In the basement, rigid foam insulation

Uninsulated basement walls can account for 20% of a home's heat conduction losses. But builders waste a lot of lumber furring out exterior basement walls just to install a layer of insulation. A more economical approach is to use 1½ in. of rigid foam insulation board, metal framing Z-bar, and drywall. Just install a pressure-treated 2x horizontally along the basement wall, then build up the surface with insulation board held in place by metal Z-bar. The metal holds the insulation in place and provides a nailing flange for drywall.

Windows, Doors, Driveways, and Walks

The exterior shell of your house, including siding, roofing, and windows, has an important task. Together, these components mediate between indoors and outdoors, admitting the right amounts of light and ventilation while blocking unwanted moisture and wind. It's a utilitarian chore, but it can be done with a little panache.

A streetscape depends as much on the color and materials of roofing and siding as it does on the overall architectural elements of the house. These choices define what real-estate agents call "curb appeal," and it ultimately influences how quickly the house will sell and, of course, how much you enjoy it yourself. Aesthetically pleasing homes can be built without elaborate trim and expensive finishes. The key is good design that uses standard elements, such as

Choosing materials for the shell of a house—including siding, windows, and doors—means finding the right balance between visual appeal and budget concerns.

Inexpensive faux brick and manufactured stone camouflage the raised entry on this home built on a flood plain.

balanced window and door placement, varied rooflines, and simple, inexpensive decorative touches such as window boxes, balconies, shutters, and decorative lighting.

Maintenance is another important consideration. Most of us want trouble-free homes we don't have to fuss over. Most would rather not have to paint. These days it is possible to build with attractive but low-cost materials, such as vinyl siding, that will last a very long time without any maintenance. Even bricks and tiled roofs can be incorporated into the design without breaking the bank—as long as you know how. Let's start at the top.

A false balcony adds old-world charm to this $145,000 home in Los Angeles, making sense of a sliding-glass door that is used to bring light and a sense of space into a small master bedroom.

Pitched Roofs Are Better

Roofing systems fall into two categories: low-slope and pitched. Although low-slope roofs may cost less to build, they require a perfect waterproofing barrier over the entire surface and at every penetration. Frank Lloyd Wright loved low-pitched roofs, but his homes became notoriously leaky. Unless you have the gumption to respond as Wright did—"Move the table," he advised when a customer called to complain of a drip over the dining-room table—you should always choose pitched roofs. This doesn't mean using dramatic slopes, but rather gentle slopes of between 4-in-12 and 6-in-12. They are easy to frame and perform well.

Roofers prefer simple designs and walkable inclines. If you make your homes easy to roof, you can afford better-quality shingles. This is how Oakwood Homes, in Denver, manages to offer concrete tiled roofs on a house that sells for $100,000. Although they could choose cheaper materials, the tile provides an important streetscape element that makes their neighborhoods appealing. If asphalt or fiberglass shingles are more common in your area, you can still build an attractive streetscape by choosing colors carefully and adding a few details.

Vinyl siding and aluminum fascia have become the most popular siding materials because they require no maintenance and are available at a reasonable cost.

Dressing up basic asphalt and fiberglass

In many areas, builders cannot use asphalt shingles because they don't carry an "A" fire rating. In other areas, builders avoid fiberglass shingles because high winds and frosty weather make them impractical to install. Fiberglass shingles don't seal readily in winter and blow off too easily in spring. An ideal blend of good fire rating and strength, plus aesthetic appeal and an affordable price, comes from laminated architectural shingles. They combine asphalt's durability with an "A" fire rating and a textured appearance that resembles natural roof products like shake. Yes, they cost a little more—but not much more—and if you have large expanses of roof showing from the street, the added expense may be worth it.

> If you make your homes easy to roof, you can afford better-quality shingles.

It's also possible to build a beautiful roof with less expensive three-tab shingles by selecting colors carefully and avoiding large stretches that show from the street. An ideal structure for three-tab shingles would be a two-story home with no more than a 5-in-12 pitched roof that runs parallel to the road and has a few high-profile gables. A decorative fascia will draw attention away from a lackluster roof covering. A color consultant can help pick a roofing, trim, and siding package that displays high aesthetic appeal even with the most inexpensive materials.

For economy, some production builders install laminated architectural shingles on roof sections facing the street and three-tab shingles of the same color in back. This creates a warranty conflict, since the thicker laminated shingles carry a 30-year warranty while the three-tabs are guaranteed for only 25 years. Yet, in a practical sense, these extended warranties mean little. Defective shingles usually fail right away.

TRADE SECRETS

On low-slope roofs and in areas of wind-driven rain and snow, you can use a layer of 36-in. rolled roofing set in vertical grade adhesive along the drip edge in lieu of more expensive rubberized ice-dam material.

Oakwood Homes uses concrete tile and roof designs with practical pitches to provide an attractive streetscape in its Morgan Park community near Denver. Homes here cost $95,000 to $145,000 in a market with a median price of $260,000.

Laminated fiberglass-asphalt shingles provide an ideal blend of affordable price, good fire rating, and aesthetic appeal.

TRADE SECRETS

Fiberglass shingles cost less than asphalt. If you live in an area with cold winters and high winds, install fiberglass shingles during the summer, when the self-sealing tabs can bond in the heat. Once bonded, fiberglass shingles resist weather well.

A decorative fascia draws attention away from a lackluster roof covering and directs it to a jazzy color palette instead.

With their variegated colors and shadow lines, metal and corrugated fiber roofing looks good on houses—not just sheds and barns.

The most common manufacturing problem you'll encounter with asphalt or fiberglass shingles is misaligned bonding tabs that prevent shingles from adhering to one another. You can fix this flaw with a caulking gun loaded with roofing cement. Most roofing problems, however, are a result of installation mistakes, not manufacturing defects. There's really no such thing as a "cheap" roofing material. Every shingle you can buy comes from a major company and carries at least a 25-year warranty. So don't fear inexpensive shingles—fear inexperienced roofers.

> Most roofing problems, however, are a result of installation mistakes, not manufacturing defects.

If three-tab shingles seem too basic, spend a bit more for variegated, open-cut three-tab fiberglass shingles from CertainTeed®. Their Hearthstead® shingle and their Patriot™ AR both provide the depth and color variation of architectural shingles at a much lower cost. They cost only about $5 more per square (100 sq. ft. of roofing) than standard three-tab shingles.

Several companies, such as Crown® and TimberTex®, make an architectural hip-and-ridge shingle that can be used to enhance a three-tab application. Manufactured with modified asphalt over a thick fiberglass mat, these enhanced hip-and-ridge shingles match most roofing colors. Apply them over the ridge or along gable ends to provide a textured look reminiscent of wood shakes while avoiding the cost of architectural shingles.

Tile without the tile price

Popular roofing materials replicate the look of natural substances like wood shakes, thatch, and clay tile. The natural materials often cost more or require specialized skills to install, but tiles made from concrete, recycled plastic, or dense, reconstituted wood fibers provide an alternative with excellent street appeal.

If you're going to use a tile roofing product, the simplicity of the roof design, more than any other factor, determines the cost of installation. Make sure the roof can handle the weight of regular tile, as opposed to lightweight tile, since the latter is only marginally easier to work with and costs substantially more.

Some sheet products, such as Ondura™ from Nuline Industries®, mimic the look of traditional Spanish tile at costs rivaling asphalt shingles. But the most cost-effective, standard product is concrete tile.

For Siding, No Maintenance Is Best

Stucco and vinyl represent the best and most popular siding alternatives available, and the reason is simple. They need no regular maintenance. Lightweight fiber-cement siding, such as Hardiplank™, also has become popular, especially where codes require fire-resistant exterior construction. But these products require paint. Although James Hardie Building Products has come out with a line of factory-stained or painted fiber-cement siding, the added costs offset any savings of not having to paint the siding once it's installed.

Depending on where you live, you may find that stucco fits within a tight budget. If not, vinyl siding is generally inexpensive in every market. In some places, a skilled labor force and a ready supply of local brick make selective use of this material affordable. Alternative masonry products are described later in this chapter.

Bonding tabs can become misaligned during a factory shingle run, and then the shingles don't bond correctly, making them more susceptible to blowing off. The repair involves squeezing roofing adhesive under every shingle lap.

Hearthstead shingles cost a few dollars more than three-tab fiberglass, but angled cuts on the tabs give deeper shadow lines that mimic the look of architectural shingles at about a third less cost.

Three-dimensional ridge shingles enhance the look of a three-tab application, making an inexpensive roof look as if it's made with architectural shakes.

Spanish Mission tiles look great, but they weigh heavily on both the roof structure and your pocketbook.

Spinning Gold from Straw

Interior designer Lolita Dirks of Englewood, Colo., caters to a unique clientele: production homebuilders. A merchandising genius who transforms insipid models into inspired homes, Dirks has become one of the nation's leading interior designers by developing strategies to create the look of high-cost details without the expense.

Dirks studies lavish details, then sketches the lines and scale of the design to tease out its essence. Once she's figured it out, she finds ways of reproducing the detail with simple stock materials and a minimum of labor. She found a way, for example, of reproducing the look of an elaborate coffered ceiling, which normally costs as much as $14,000, for $100 in materials and labor. Here are a few of her tips:

> She finds ways of reproducing the detail with simple stock materials and a minimum of labor.

- For ceilings, try using wood trim, crown molding, or plastic in geometric shapes to create interest and intricacy. Use lighting to enlarge a space or make it seem more intimate.
- Don't be afraid to draw on walls with wood trim, layered drywall, and other inexpensive materials to frame art or create features like a headboard for a bed.
- Create dimensions with mirrors: Mirrors can augment a small space or reflect beautiful outdoor scenery at very little cost.
- Accent with paint. A single wall painted with deep tones or a faux finish can turn an ordinary room into something whimsical or elegant.
- Experiment with nontraditional tile on the fireplace surround. Try mosaics instead of the same old 8x8s and 12x12s seen so often. Buy a granite or marble remnant for the hearth, and don't bother finishing the edges.

Inexpensive, MDF bead molding, accent paint, and mirror tiles add class to an otherwise down-to-earth dining room conceived by Lolita Dirks.

Fact Sheet

Who: Lolita Dirks

Where: Englewood, Colo.

What she does: A lot with a little. Using imagination and inexpensive materials, this interior decorator specializes in creating details that look much more expensive than they are.

Met-Tile steel panels and Nuline's corrugated fiber roofing can be installed over standard roof framing without reinforcement. They are more expensive than asphalt or fiberglass shingles, but a lot less than tile.

Brick and vinyl siding, both no-maintenance exterior finishes, can be combined on the same house.

Combining sheathing and siding

The least expensive siding products include one-step sheathing and siding panels like T-1-11. In areas where shear requirements demand extensive solid sheathing, these panels can save hundreds of dollars. Many wall-panel manufacturers will install preprimed T-1-11 siding on exterior walls, which greatly speeds completion of the building's exterior. In this instance, an option for windows is to use the kind without a flange (retrofit, remodeling windows) and attach them through the frame. Then case around the exterior with brick mold or 1x4 material.

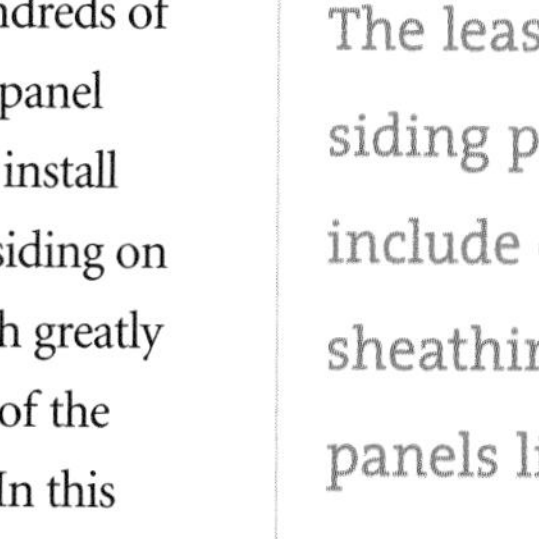

Some sheathing and siding panels install horizontally to mimic the appearance of lap siding. Other panels have vertical grooves milled in the face to imitate traditional bat-on-board siding. Even when factory-primed, all of these panels require a coat of paint and regular maintenance. Still, siding and sheathing panels represent the lowest cost siding alternative. Used with architectural skill, they can provide an attractive, solid exterior surface that can be upgraded later.

Alternatives to lap siding and shingles

Cedar boards and shingles are among the most widely regarded exterior surfaces. Buyers like the natural, classic look of wood siding, but the price and problems associated with high-quality natural materials have relegated them to less than 3% of the siding market. Instead, homebuyers and builders have turned toward manufactured products that generally cost less than wood and provide advantages in installation, durability, and maintenance. The most popular of these is vinyl.

TRADE SECRETS

If you use vinyl siding over 24-in. on-center framing, sheath the entire house with OSB to prevent the siding from buckling. In areas where sheathing is not required, space your exterior studs at 16 in. on center.

San Francisco architect Donald MacDonald used simple, exterior-grade plywood for siding on his whimsical and highly affordable homes.

A type of plywood siding called T-1-11 is an affordable option for finishing the outside of the house.

Vinyl is durable and affordable Vinyl siding has become the most popular version of traditional lap siding because it's affordable, durable, and maintenance-free. Newer vinyl products resist waviness better than the older versions and remain colorfast virtually forever. When you subtract the cost of painting, vinyl siding is the least expensive exterior cladding available. It comes in a wide range of colors, textures, and lap styles. Quality ranges from a 3-mil builder grade to deluxe 5-mil panels with proprietary stiffening elements to prevent waviness.

A variety of trim—from corners to soffits and fascia—is available to match every color and style. In cold climates, vinyl can be installed during the winter whenever temperatures climb above freezing. Not having to paint makes it easier to close up a house quickly in almost any climate.

Hardboard siding has improved Hardboard siding earned a bad reputation in the wake of widely published reports of failures in the early 1990s, but manufacturers have developed new products with better adhesives and chemical treatments to prevent swelling and rot. Manufacturers have also developed hi-tech descriptions for their siding panels, such as Georgia Pacific's high-resin, high-temperature, all-wood fiber composite, which helps hide the identity of their products. But modern hardboard products come with long warranties, ranging from 10 to 30 years, and manufacturers appear confident in these redesigned products.

Primed hardboard soffits and fascia can be combined with vinyl siding. Especially at the rakes, hardboard soffits are faster to install than vinyl ones. Hardboard doesn't blow off or mar as easily as vinyl or aluminum cladding. When damaged, it can be touched up with paint.

After you factor paint into the overall cost of hardboard, vinyl still costs less. But where architectural restrictions prevent the use of vinyl, hardboard provides a next-best alternative with excellent durability, paint adhesion, and warranties. To prevent water buildup behind the siding, it's a good idea to use a ribbed house wrap, such as Tyvek StuccoWrap™, which provides an easy-to-install drainage plane.

Vinyl panels come in a broad range of styles and prices. To avoid overspending, stick with stock profiles and simple moldings. Skip the designer trim and accessories.

Brick facades help to promote sales of NuHome's affordable communities in Houston. Prices range from $95,000 to $125,000.

Hardiplank provides a competitive fire-resistant alternative to vinyl siding.

Fiber cement provides fire resistance Since James Hardie introduced fiber-cement siding, it has become the most popular alternative to vinyl. Fiber cement installs like traditional wood or hardboard, but it has some of the advantages of masonry. It holds paint on a par with the old asbestos siding, which is to say very well, and resists fire. And although it's meant to be painted, you could leave it unprotected—the 50-year warranty still holds.

> Since brick and stone represent the most costly exterior finishes, builders use these materials primarily on accent walls.

Trim, soffit, and fascia materials are available to make a 100% fiber-cement exterior. Although lap siding remains the most popular application, fiber-cement products are also designed to resemble shingles, stucco, brick, and stone. James Hardie produces a line of factory-painted and -stained boards that eliminate the need for site finishing. However, these generally require great care to install and extensive touch-up.

From a cost perspective, fiber cement provides an excellent alternative for decorative exterior finishes where codes require fire-resistant cladding. Nichiha® Wall Systems sells an excellent imitation brick or stone panel that offers the feel and appearance of masonry without its weight and cost. The panels install with a proprietary mounting system that holds them off the sheathing, just like brick. These panels can be used as feature walls or as cladding for the entire house. They cost more than other types of siding but less than masonry. They can be installed in winter, when it's too cold and cumbersome to lay brick.

Brick and stone are the most expensive options There's something about masonry that spells home. But since brick and stone represent the most costly exterior finishes, builders use these materials primarily on accent walls. Other choices that look very similar can help reduce costs.

Nichiha Wall Systems makes panels that look like real masonry but cost less, weigh less, and can be installed even in freezing temperatures.

Natural stone is an appealing but often expensive material—unless you happen to live in a part of the country where supplies are abundant. It's worth checking.

The most common natural stone is fieldstone. There are many varieties that may be available locally and inexpensively by the ton.

Fake stone that looks real Builders have turned to artificial stone almost to the exclusion of the natural substance. Light, factory-made stones resemble the real thing, are easily applied, and generally cost less than quarried stone or fieldstone—but not always. If there is abundant natural stone in your area, you may find that one or two local types of stone actually cost less. Even when natural stone costs more, masons often charge less to install it. Masons prefer laying the real thing because they don't have to install the lath and scratch coat required with manufactured stone.

Even when natural stone costs more, masons often charge less to install it.

Of course, artificial stone provides variety and unlimited supply almost anywhere in the United States. Unlike natural stone, the manufactured products don't require a footing and mount on virtually any solidly built surface. Over sheathing or studs, you must install building paper, wire lath, and a scratch coat before laying up the stones. On a clean block or concrete wall, the stones can simply be mortared in place.

When using manufactured stone, you'll find that the corners and trim pieces add substantially to the cost of a basic installation. Many builders opt to trim corners with wood. Some of them just butt plain stones together with grout.

Although manufactured stone corners add depth and realism, you can save money by simply butting plain stones at outside corners.

Brick is often available locally Brick remains readily available and inexpensive, at least compared with stone. Most areas have a brick plant for local supply and many skilled masons competing for work. Averaging at about $3 per square foot for materials and labor, brick affords a lavish look for only about twice the price of vinyl siding. King-size or oversize brick represents the most economical alternative because it only takes four-and-a-half bricks to fill a square foot of wall, as opposed to the standard modular brick, which requires seven. Masons generally charge more per unit when using oversize brick because they want to protect their square-foot price. But if you negotiate skillfully, you should be able to get a cost break in labor, as well as materials, because fewer bricks require less time to complete a wall.

You can purchase brick for about a third less by buying factory overruns and color defects. Make sure you buy enough to finish the job, however, since you may not be able to buy any more that match.

Although split bricks cost more than full bricks, practically anyone can install them. Some split bricks come mounted on wire mesh so installation amounts to spreading thin-set mortar on the wall and pressing the brick and mesh into place. Finish the installation by mortaring the joints with a grout bag and brushing the lines clean when the mortar starts to set up.

Since the most expensive aspect of any masonry product is the labor to install it, consider doing it yourself with NovaBrik™. A decades-old Canadian product relatively new in the United States, NovaBrik is a dry-stack, siding-like lapped concrete brick that provides the full dimension of masonry at reduced cost and with easy, low-skill installation. NovaBrik is attached to the framing with structural screws, requiring no brick ledge or mortar. It comes in different color palettes so that you can create your own blend. The manufacturer provides corner blocks, wainscot, and windowsills. The installation literature and instruction video provide excellent training, so your first project should turn out right. Although on close inspection, the lack of mortar joints reveals you're not using real brick, this product has a handsome quality of its own that does not look like an imitation of something else.

NovaBrik has created a lapped, dry-stack concrete brick that features the full dimension of masonry at reduced cost and easy, low-skill installation. NovaBrik stacks like siding with screws, requiring no brick ledge or mortar.

Brick is often readily available at prices lower than natural stone. At this Lincoln, Neb., factory, clay is extruded in a rectangular ribbon, then cut to size and fired.

Stucco makes a durable, low-maintenance exterior finish and may be a bargain in regions where skilled plasterers are plentiful.

Stucco is durable and low-maintenance

In cities where skilled plasterers abound, three-coat concrete stucco provides an easy, inexpensive means of building a maintenance-free exterior finish. Stucco can actually be used over a framed relief to create window trim, cornices, soffits, and even fascia. The plastic quality of the material makes stucco an ideal choice for siding an entire house, with accents built into framing at streetscape elevations.

Because stucco receives paint readily, color accents can be added inexpensively for a highly customized look. Stucco tools easily, too, so you can scribe patterns into the surface to define wide moldings around windows and doors, corner blocks, and even a realistic imitation of brick.

Synthetic stucco can cost 50% less than three-coat concrete in areas with few skilled plasterers, and hence little price competition. Unfortunately, the synthetic finishes do not provide the same solid durability of concrete and are prone to warranty claims. If concrete stucco is not a realistic choice, an alternative is using fiber-cement boards and decorative battens to build a faux stucco accent wall.

You can carve stucco to create inexpensive but handsome faux moldings around architectural accents.

Red color-coat stucco carved to look like brick provides a convincing finish on this wood-framed chimney.

Frame an opening with a 2x4 and then plaster over it to dress up an otherwise plain window.

Windows, even relatively inexpensive vinyl units, can represent a substantial part of the cost of the exterior shell. The way to save is to reduce the number of windows in the plan.

TRADE SECRETS

Profiled grills that are factory-installed between the glass cost less than interior grills and are never lost or broken. To save money, install these internal-grill windows on the street side of the house and use plain glass in other areas.

Economizing with Doors and Windows

The best way to save money on doors and windows is to use fewer of them. You can eliminate double windows in bedrooms by setting a single window near an outside corner. In living rooms and kitchens, you can reduce the number of windows without making the room appear dark or claustrophobic by placing openings on optimal sight lines that permit plenty of light and lead the eye outdoors. Bright, reflective surfaces, such as off-white walls and pure white ceilings, can add luminosity. Windows can face east and west to maximize daylight exposure. Unless you build in a hot climate, avoid shading your windows with trees, architectural features, or other obstructions. If you can't maximize window exposure, consider a small skylight with a light well.

The best way to save money on doors and windows is to use fewer of them.

Carefully placed windows and a distinctive entry make your home more attractive from the street. But you can forgo fancy doors and picture windows on the three remaining sides of the house. As long as windows meet light, ventilation, and egress requirements, and doors provide convenient access to the yard and garage, using standard-sized, readily available door and window units will save money.

A door can serve double duty when it includes a windowpane. Sliding-glass doors provide the function of door and picture window at the same time, although sliders are prone to malfunction and may cost more than a simple, full-light swinging door. You can achieve the look of a double French door set by framing a fixed, full-light door and a swinging door next to each other.

Whenever possible, use a series of conventional opening sizes, such as 2 ft. by 4 ft., in sequence to achieve architectural effects. Three of these windows framed into a single opening with two double studs acting as framed mullions costs less than one 6-ft. by 4-ft. window yet presents a similar appearance.

Simple horizontal sliding windows are the least expensive kind, followed by single- and double-hung windows.

TRADE SECRETS

Ask your supplier to ship windows without screens and then order the screens just before finishing the house. This way, you won't waste money replacing missing and broken screens.

MORE LIGHT FOR LESS

A small skylight does not admit much light, but by framing a light well that widens at the bottom, light will be dispersed. This small, inexpensive opening can brighten up a shady room or gloomy hallway effectively.

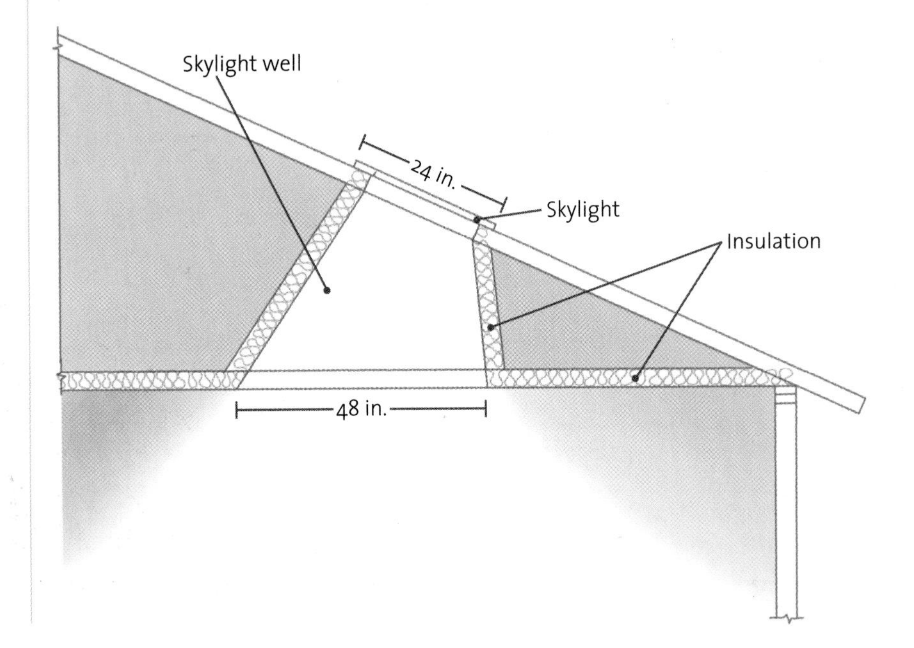

Windows with internal grills can be used on the street side of the house, and windows with plain glazing used elsewhere.

TRADE SECRETS

When installing vinyl windows, don't just check for square; use a level or other straightedge to check for bowing, too. Make sure to shim the sill, since even a sill that's slightly bowed can cause water to puddle away from the weep holes.

Low-e windows boost energy efficiency

Low-e glazing will contribute to your home's energy efficiency, and it may help reduce the size of heating and cooling equipment, at a relatively small additional cost. In areas of extreme summer heat gain, such as the Southwest, consider using standard low-e windows combined with architectural shading instead of pricey, high-performance windows designed for southern climates.

Vinyl windows have gained popularity in every region of the country. In a hot, sunny environment, it's best to avoid dark-colored frames because they absorb heat. Vinyl tends to expand and contract with temperature changes while glass does not. Dark vinyl colors in high-heat environments can warp, break the glass-to-frame seal, and start leaking. Manufacturers use special additives and reinforcement to ensure the dimensional stability of their windows. But unless you buy a very high-quality window, it's best to stick with white.

On large windows, 6 ft. wide or more, avoid cheaply made vinyl or aluminum frames that tend to warp under their own weight. It's easy to rack these soft-framed widows during installation, so check them very carefully. Even if you install the window perfectly, it's easy to bow the frame while pushing insulation into the gap between the window and the opening. Most vinyl and aluminum window leaks come from installation defects, which carry no factory warranty.

Prehung steel doors are low in cost

Prehung doors save time and cost less than doors hung on site. Steel entry doors are the least expensive. Many people prefer embossed, panel-style entry doors to flat doors. They come in two price ranges, corresponding to 24-ga. or 25-ga. steel cladding.

More expensive doors with a thicker steel skin resist job-site abuse better than the thin cladding. But unless you're extremely careful, any door will need repair by the time construction is over. Some lumberyards offer "bang" doors, which are door jambs fitted with temporary construction doors installed instead of the actual doors you purchased. Once you're ready to install hardware and finish construction, the lumberyard swaps out their temporary door for a permanent one. Your yard may not offer this service, but it's worth asking about.

You can buy vinyl-sliding and atrium-style doors for about the same or just a little more than aluminum ones. The thicker vinyl frame looks better, and because there's no need to add trim, it saves money on casing. It's also worthwhile to consider buying a smaller patio door. A 72-in.-wide door may be standard, but a 60-in. door costs about $50 less. The size disparity generally makes no appreciable difference, especially in smaller homes where the 60-in. door actually looks more appropriate.

If intense sunlight is a concern, consider architectural shading and standard low-e glazing instead of pricey high-performance windows.

SKIP THE EXPENSIVE PICTURE WINDOW

Instead of buying an expensive picture window, consider combining two or three stock windows in tandem. In this example, the center window is fixed and the flanking windows open and close. By framing the perimeter with a thicker band of trim than the material used to cover the mullions, the three windows appear as a single unit. Here, three 2-ft. by 4-ft. windows replace one 6-ft. by 4-ft. unit at a lower cost.

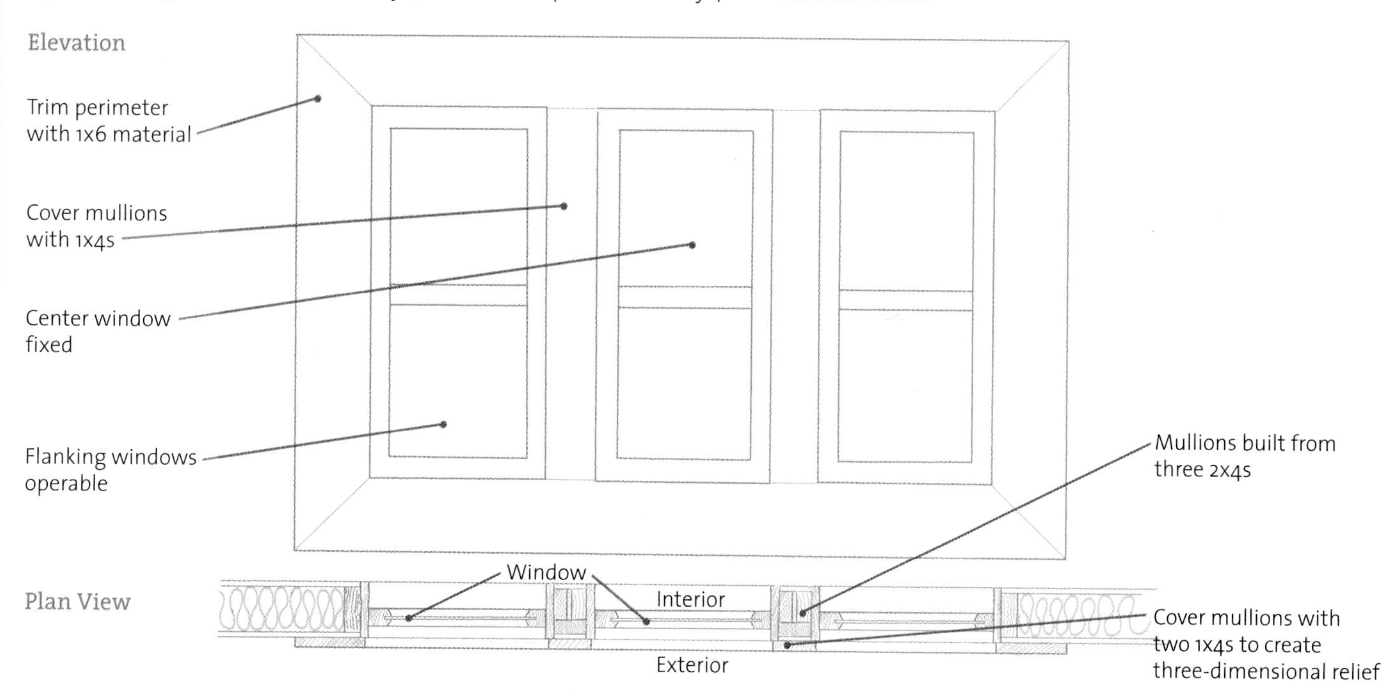

Six Inches Can Cost $100

When using overhead steel sectional doors, any deviation in the rough opening can add substantially to the price of the door. For example, if you build your header at 90 in. instead of 84 in., the additional half-foot can add more than $100 to the price of the door. Check rough opening sizes with your overhead door supplier as you would your window manufacturer. Proper framing allows installers to get in and out quickly, making them more willing to give you a lower price.

Creativity pays on overhead doors

The most expensive door in many houses actually detracts from the home's appearance. Most overhead garage doors look like a barn entrance, without the bucolic charm. In Victorian days, the transportation entrance—which back then led to a carriage house or stable—closely mimicked the architecture of the main home. Often, the carriage house had an elaborate door that enhanced the structure. Our overhead doors have become entirely pragmatic, lacking any sex appeal. But you can achieve a customized design inexpensively by using a plain plywood door and a little creativity.

False posts ripped from ⅜-in. plywood, layered boards, rail and stile framework, or even a split look that creates the illusion of two single-stall openings can actually create an architectural accent instead of an eyesore. Unfortunately, most builders opt for a factory-finished steel sectional.

Sectional steel doors provide a convenient, durable alternative to the traditional plywood garage door, albeit not a very attractive one. You can paint these doors to match your siding or leave the white, almond, or sandstone factory coatings exposed. The plain, 25-ga. wood-grained, raised-panel design costs a little less than a site-built wood door, but it lacks the architectural potential. Most overhead door companies also manufacture an upgrade line of handsome, carriage-house style doors that require no maintenance. They cost about 50% more than the standard line, but when you consider how much of your house facade is garage door, the option is worth considering.

A simple, painted plywood garage door can add to instead of detract from the architectural appearance of an affordable home. To get the same custom look with a steel door, you have to pay hundreds of dollars more.

Overhead-door manufacturers such as Clopay® have introduced insulated steel doors with a vinyl overlay that look like wood carriage-house doors. The doors are low-maintenance and low in cost.

TRADE SECRETS

For a truly bang-proof door at a bargain price, buy Therma-Tru's Smooth Star® fiberglass flat doors in flush- or raised-panel designs. These doors have the same resistance as the embossed fiberglass doors, but at a more competitive price.

Concrete driveways and walks have become commonplace, but they are not always the most affordable or sensible option.

Dressing Up Plain Concrete

When you install concrete sidewalks, you can get the look of stamped concrete without the price. Using standard masonry joint tools, just sculpt your pavement to look like tile or stone.

After edging and troweling the concrete to get the surface finish you want, use a pointing trowel to draw a free-hand scalloped or tile pattern on the concrete. Next, go over the lines you've drawn with a small groover to create a pleasing border on the pattern (1, 2). Make sure to include a groove wherever you would place a control. Then go back with a margin trowel (3) to cut through the concrete deep enough to force shrinkage cracks to occur where you want them. The result is a surface with a lot more variety and visual appeal than a plain concrete walk (4). For tiled squares, the technique is the same. Just use a 1x4 as a straightedge (5).

For Walks and Driveways, Skip the Concrete

Concrete walkways, stoops, and driveways have become a standard feature of modern construction. Twenty-five years ago, it was not uncommon for builders to provide a small wooden porch and steps, stepping-stone sidewalks, and a gravel driveway. With our new concern for environmental sensibility in construction, those days are coming back.

Using permeable materials for driveways, parking lots, sidewalks, and even streets can minimize storm-water runoff, which has become a major environmental problem in urban areas. Excess runoff caused by too many acres of roofs and paved surfaces contributes to floods, river erosion, overtaxed storm-water systems, and pollution. Substituting less costly materials like gravel for concrete won't set you apart as a cheapskate, but an environmentally sensitive homeowner or builder.

Rock driveways are cheaper

Besides saving you thousands of dollars, a well-constructed rock driveway provides benefits that concrete cannot. In colder climates, a rock drive offers a slip-resistant surface year-round. Frost heaves, de-icing compounds, and

Driveways constructed with crushed rock may cost thousands less than concrete or asphalt, but cost isn't their only advantage. They also perform better in icy conditions and provide a permeable surface that helps control runoff.

BUILDING A SIMPLE DRIVEWAY

There's no rule that driveways must be poured concrete or asphalt—a stone drive is a low-tech alternative that may end up costing less. For a temporary access drive, spread a layer of 3-in. rock over the future driveway at the start of construction. When you're ready to finish the driveway, install road fabric and add a layer of crushed stone followed by road gravel for a smooth, durable finish.

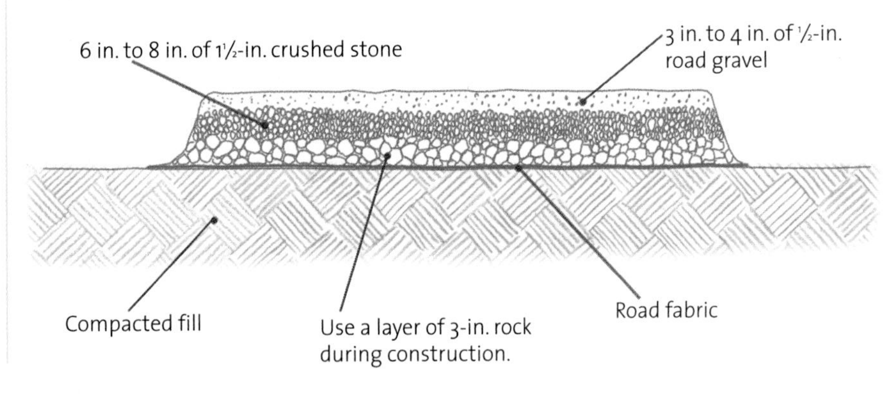

expansive soils do not damage it. You can build a gravel driveway in freezing weather. Gravel doesn't block drainage patterns or create runoff problems, so driveways made from it do not require expensive drainage systems. After construction is complete, it's easy to renew a rock surface with a fresh load of inexpensive stone and a rake.

Asphalt paving is a good value

When you have to pave your driveway with a permanent material, consider asphalt. The asphalt mix used for paving residential streets is generally the least expensive and most readily available kind. On a properly prepared subgrade, asphalt paving can last 20 years and provide excellent results at approximately $1 less per square foot than concrete.

Whether using asphalt or concrete, always consider the size and length of the driveway. Avoid placing the house any farther back from the street than required by zoning regulations, and pave a driveway no wider than the garage door. It may even be possible to reduce the width of the driveway as it approaches the street. This can save a substantial amount of money since curbs and city-inspected approaches are usually much more expensive to build than a private driveway. Funneling a 16-ft. driveway into a 12-ft. approach generally saves between $200 and $300.

Paving with soil cement

To get ideas on economical sidewalks and driveways, look through a few gardening magazines—you won't see much concrete. In fact, stamped concrete is just a high-end imitation of the natural and inexpensive walkways we used to see.

Meandering paths made from flagstone set over compacted ground with creeping thyme growing between the stones; crushed rock delineated by wood edging; stepping blocks made from recycled timbers set in gravel—these are the kinds of sidewalks our homes had before concrete became the most popular choice for paving. They look even better than stamped concrete and harken back to a quieter, more tranquil time.

One of my favorite techniques for creating low-cost private sidewalks involves paving with "soil cement." This is a blend of natural soil with inexpensive, low-grade cement to create a compact, dust-free walking surface. If your site contains sandy or coarse low-silt ground, just till a 4-in.-deep path. Remove any organic matter, like roots and weeds, and then form the edges of your path with treated wood. Spread about a sack of dry cement over every 40 sq. ft. of surface and work this cement into the soil with a tiller. Strike off the path with a screed board, as you would concrete, and then tamp the surface until it is firm. You may need to add another layer of soil and cement to bring the path back to level.

When the surface looks the way you want it to, wet the soil cement with a fine spray. Let the moisture soak into the ground and wet it down again. Once the surface has dried so that it's not sticky, use a roller to smooth the path. Keep your sidewalk covered with plastic for a few days and don't walk on it until it cures, about one week.

If you do decide to go with concrete walkways, there's a fairly simple method for sculpting the surface and dressing it up, rivaling the more expensive stamped concrete process (see the sidebar "Dressing Up Plain Concrete" on p. 173).

Money-Saving "Green" Checklist for Siding

- Hardboard is made with recycled wood fiber, which won't split or warp but will hold paint beautifully.
- Fiber-cement siding, guaranteed for 50 years—even if you don't paint—is certainly more durable than wood and reduces demand for forest products.
- ArgiStain is an inexpensive soy-based, nontoxic penetrating stain suitable for interior and exterior concrete floors or masonry surfaces, such as brick.

The best looking residential sidewalks are not concrete. Consider using paving stones as a do-it-yourself solution that looks better and costs less than calling a contractor.

BUILDER'S CORNER

Appearances Really Count

For builders concentrating on the affordable-housing market, it's all about sex appeal and the bottom line. You can try educating your buyers into accepting a well-built, but somewhat boring house for practical reasons—or just make your homes more attractive. Aesthetics play a vital role in human nesting, and you should seek to please rather than change your buyer's tastes. Affordable-home builders always struggle with balancing budgets and beauty: When your homes are appealing and don't make the streetscape look cheap, you protect your reputation in the community and the ultimate success of your company.

10 Interior Finish

Building codes establish a baseline for many construction costs, and as a result builders often consign the bulk of their cost-cutting efforts to the finish trades. Here, plenty of potential exists to make thrifty choices, such as using low-end cabinets and inexpensive carpeting. This is precisely where value engineering can be the most frustrating. It too often seems that building affordably means reducing value rather than maximizing it.

One way of avoiding this trap is to devise an "options menu" from which to choose features that make a big impact while skipping other upgrades that aren't as important. Having too many options can cause job delays and ultimately raise the cost of building a house. For example, it's easy to change door styles in a kitchen, but it's problematic and more expensive to

Interior finish work, including the installation of doors, trim, and floor coverings, presents a variety of opportunities to cut costs.

BUILDER'S CORNER

Keep the Project Moving

Devise a list of options for homebuyers that will cause the fewest disruptions to the construction schedule and allow the project to move forward without delays. Here are a few suggestions:

> Offer options on cabinet doors in the kitchen, but not on cabinet boxes.
> Keep flooring options within specified types, such as carpet and vinyl.
> Limit countertop color and pattern choices to the stock you have on hand.
> Offer a broad color selection for accent items only, such as the exterior face of the front door, which you can paint right before the move-in date and which takes one quart of paint.

Options like revised cabinet layouts and floor-plan revisions tend to throw off the production schedule and invite subcontractor errors. Even when you charge for changes, it's rare that you'll earn enough to make up for the trouble caused by customizing a production plan.

This fancy fireplace was made entirely of drywall and paint.

revise the cabinet layout. For builders, this means giving buyers a menu of competitively priced door options, but only restricted layout choices. In this way, the base price of a house can remain low while buyers have a measure of autonomy.

Design One Detail at a Time

One way of adding value to interiors is to develop a scrapbook of million-dollar home details that can be reproduced economically. Sometimes all you need is a detail or two to create the perception of luxury without the cost. An archway, a bullnose corner, or a ceiling coffer built from lumber scraps and drywall can make an inexpensive home look posh. You can mine ideas from decorator books and home shows on television.

Large home-building companies usually enlist decorators to come up with creative merchandizing concepts for their model homes. You can take advantage of this design bounty by touring models whenever possible. A single, innovative detail can make a big difference.

Homebuyers tend to value ineffable qualities like "whimsical" and "quaint," while they assume and expect "well-built," "safe," and "durable." Houses need a little sex appeal.

Add light to the interior

Two of the most valued interior details, light and air, come absolutely free. A little extra headroom with a higher ceiling and windows in unexpected places add beauty to a house. For example, a large window in the kitchen can take the place of several feet of more costly upper cabinets. The cabinets won't be missed because the extra light is more enjoyable than a bit of extra storage. Likewise, if the house is sited carefully to exploit exterior views, the interior becomes spectacular without any upgrades at all. A tree enhances not only the yard, but also any room with a window overlooking it. In the absence of natural features, consider a trellis with a flowering vine right outside a small, decorative window in the kitchen or dining room.

You can usually find a place for almost-free framed niches that inspire creativity.

Using a framed wall wrapped with drywall and bullnose corners, a Santa Fe, N.M., builder adds a kitchen peninsula using inexpensive stock cabinets.

Vaulted ceilings add cost, but they create the illusion of openness in otherwise small spaces. Adding a few inches of volume costs less than a few square feet of floor.

In this kitchen, the picture window may seem like an added expense, but it actually costs less than a row of cabinets.

You may be tempted to leave the installation of drywall to a subcontractor and simply pay the going rate, but there are ways to make this common wall finish more affordable.

Making the Most of Drywall

Drywall originally replaced wood lath as a base for plaster, hence the term plasterboard. It became a stand-alone product after about 50 years, when joint compounds and taping systems evolved. A remnant of plasterboard's early history persists in "thin-wall" applications, where sheets of drywall with joints taped with fiberglass mesh become the substrate for a single coat of hand-troweled finish plaster, an excellent and economical way of finishing a wall in one step.

Although builders generally relegate installation decisions to their subcontractors and then pay the going square-foot rate, there are ways to reduce the cost of both installation and finishing, making drywall even more affordable than it is already. Detailing helps, too. Drywall arches, for example, once were reserved for high-end housing because of the expensive, time-consuming framing they required. Now anyone can buy a preformed arch that installs in minutes.

Take advantage of the variety of lengths that drywall comes in to reduce the number of joints. That saves finishing time and money.

Reducing the cost of hanging drywall

Joints, corners, and nails represent the three expense points in drywall installation. By reducing these, you reduce costs. This is why drywall comes in lengths of 8 ft., 10 ft., 12 ft., and 14 ft. When full sheets are used wherever possible, the number of joints on a wall can be reduced. Where joints occur, instead of cutting two sheets to meet over a stud, consider using joint clips that allow you to float drywall over an open bay.

If you plan to panel a wall, instead of using standard tapered-edged drywall with joints taped, use tongue-and-groove sheets for a no-tape finish.

To reduce the number of fasteners required, use drywall adhesive and simply pin drywall in place with screws to set the mastic. This may not reduce costs significantly during installation, but it certainly streamlines the finishing process. Adhesives work especially well when applying a double layer of drywall at a firewall. You can actually glue and then pin the second layer onto the first with a few double-

The old way of creating a drywall arch was to frame the opening with plywood or OSB, add blocking, and then skin the arch with drywall. It's a slow process.

Premade arches have made it simple and cost-effective to add this attractive detail to door openings. Arches are nailed in place quickly, and the jamb skinned with ¼-in. drywall.

DRYWALL JOINTS MADE EASY

By using Prest-On drywall joint clips, there's no need to lap drywall joints over studs. This may allow the use of longer sheets, reducing the overall number of joints that must be taped and finished and cutting costs.

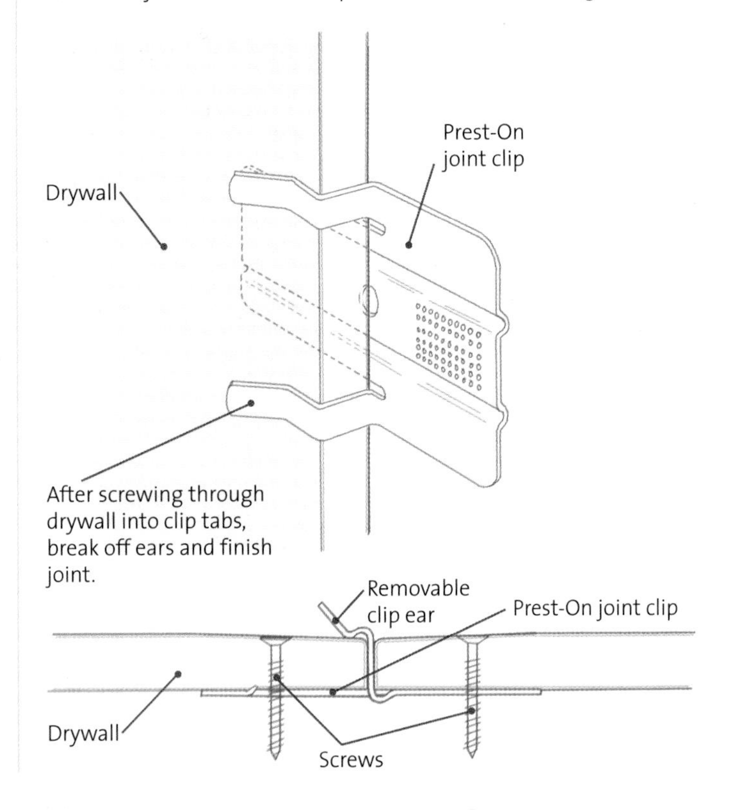

Money-Saving "Green" Checklist for Interior Finish

> MDF moldings and panels offer a very low-cost green-building alternative.

> Choose light interior colors for walls, ceilings, and floors. These help reflect natural light and reduce the need for artificial lighting.

> Recycled-content carpet is readily available as PET (polyethylene terephthalate) polyester. It's generally made from high-quality food-compatible products such as soda bottles and milk containers. You can also buy recycled cotton and wool carpets.

> Natural linoleum is manufactured from entirely benign products like cork and linseed oil. It contains no petroleum products or chlorinated chemicals. Yet, linoleum is more resistant to tearing and scuffing than vinyl and usually lasts about 40 years.

> Bamboo, a fast-growing grass that yields a hard, stable plank, can be used in place of hardwood flooring for about $4.00 per square foot.

> For slab-on-grade construction, one of the least expensive flooring solutions is to saw-cut, stain, and polish the slab as a hard-finish floor surface.

> ArgiStain is an inexpensive soy-based, nontoxic penetrating stain suitable for interior and exterior concrete floors or masonry surfaces, such as brick.

headed nails. Remove them in 24 hours, after the glue sets, and then finish.

Although nails and screws have become the preferred fastening systems in residential construction, commercial contractors also use pneumatic staples for speed when hanging double-layer installations and firewalls. Wide-crown framing staples can be used to install the firewall layer on townhouse and garage walls.

Bulkheads and cabinet soffits can be built easily by borrowing metal-stud techniques from commercial drywall operations. Instead of framing a box and laminating it with drywall, commercial installations use the drywall itself as a structural element.

Using drywall imaginatively

When installed correctly, water-resistant drywall can be used with confidence under ceramic tile or other bathroom finishes. It's less expensive than substrates such as concrete board and cement plaster. To avoid problems, make sure to leave a ½-in. gap between the board and bathtub or shower pan. Coat all the edges cut for pipes with a water-resistant sealant and size the surface with waterproof mastic before gluing on tile.

DensShield®, often called "grayboard," is a gypsum product that provides another lightweight, inexpensive alternative to standard tile backing, but unlike water-resistant drywall, it can actually be submerged in water without damage. Most builders limit their use of drywall to the interior, but exterior-grade drywall trimmed with wood batts can provide an inexpensive soffit or porch ceiling. Type X exterior drywall also provides an inexpensive fire-rated sheathing in jurisdictions where exterior surfaces must meet certain fire codes. Instead of sheathing with OSB and then covering your walls with a fire-resistant material, use Type X exterior drywall overlaid with an inexpensive siding such as vinyl.

Textured surfaces and speedy corners

Along the Pacific coast and in the Southwest, textured walls have become commonplace. In other areas, they're not. But coast-to-coast, orange-peel and knockdown textures cost less than a smooth finish while adding a three-dimensional quality that can enhance the interior. A textured finish also camouflages bulges, bruises, and other common drywall defects.

When installed correctly, water-resistant drywall can be used with confidence under ceramic tile or other bathroom finishes.

Bullnose corners remain popular in many markets but generally cost a little more than standard square corners. Drywall Systems®, creator of the No-Coat® drywall corner, manufactures a bullnose paper tape that sets in mud and finishes with a single coat of joint compound, reducing labor.

Drywall Systems also makes interior and exterior paper corners and flex bead. Their original No-Coat drywall corner has a tapered plastic membrane that provides a stiff, durable corner with a one-step operation. Again, this eliminates the need for metal bead and a three-coat finish. The company's UltraFlex® paper corner provides an excellent means of finishing long peaks on vaulted ceilings, where standard joints tend to crack as wood trusses dry and separate.

USING WATER-RESISTANT DRYWALL

Water-resistant gypsum drywall can be used safely as a tile substrate in the bathroom, providing it is installed correctly. It's less expensive than concrete backer board.

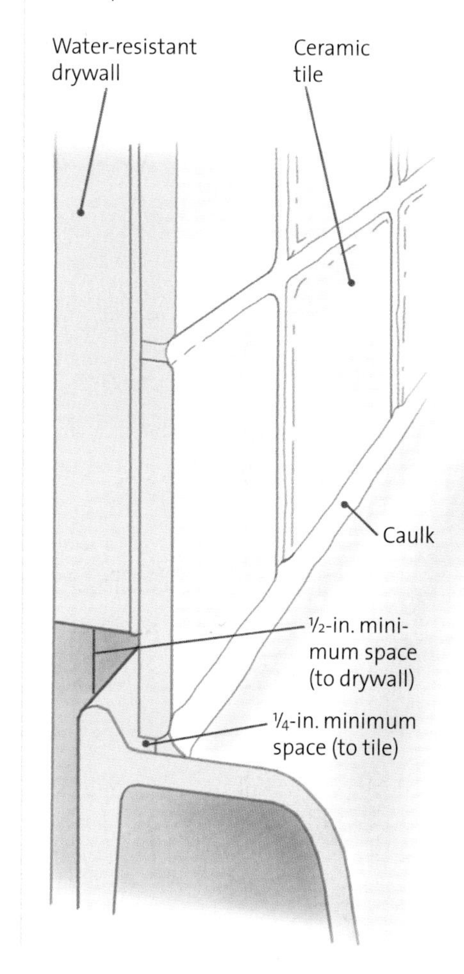

SIMPLE DRYWALL SOFFITS

Soffits, like those over a set of upper kitchen cabinets, can be made quickly with nothing more than drywall, furring channels, and corner bead. It's less expensive than making them from conventional framing materials and then adding drywall.

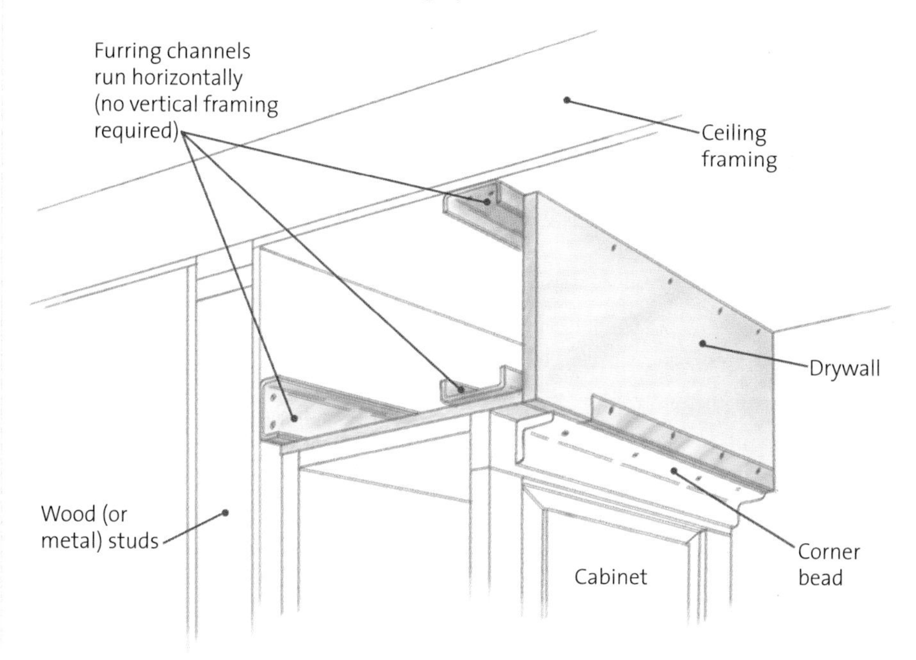

Textured surfaces on drywall reduce costs by eliminating one finishing step.

No-Coat bullnose is applied with a proprietary tool. This type of corner is much faster to install than conventional metal corner bead because it requires fewer coats of joint compound.

Choose Paint Carefully

Water-based paints have become universally accepted on interior walls and are used increasingly for woodwork. Over clean drywall, water-based paint can be applied without a primer or sealer. Flat, off-white colors conceal minor irregularities in walls and ceilings and provide a luminous surface that enlarges a room's appearance.

> Over clean drywall, water-based paint can be applied without a primer or sealer.

You can use semigloss water-based enamels for a hard, scrubbable finish on woodwork and kitchen walls, but many builders economize by painting baseboards and walls with the same flat coating. While this practice may not work well on raw wood, it's perfectly adequate for preprimed and MDF products.

Whenever possible, avoid painting the ceiling. A textured drywall ceiling saves considerable time in painting a surface that shows many imperfections when smooth. If you paint ceilings, spray them before installing the woodwork and roll the walls after trimming the house.

Spraying paint is not always a money-saver

At some point, most of us reason that spray-painting the interior of a home saves time and money—but experience teaches otherwise. Once you add the time required for careful masking and the gallons of extra paint consumed in spraying the interior of a home, you realize there's hardly any benefit over rolling the walls. Touching up spray-painted walls can be difficult, too, since brushes and rollers leave a discernable texture.

Although spraying walls doesn't save material or labor, spraying doors can provide cost benefits and a better finish. Consider using a low-volume air gun with an alkyd-based paint for this purpose, since it allows better control and wastes less material than an airless sprayer. When painting doors, it's best to use primed, paintable hinges, so that you can coat the doors, jambs, and casings all in one operation.

Instead of masking around openings, spray doors before rolling the walls. It's easier to cut walls to trim than it is the other way around. Especially if you use the same color on walls and woodwork, you can cut without worrying about drawing perfect edges. Matching trim and wall colors helps to de-emphasize the trim for a less cluttered appearance, making smaller rooms appear large.

Keep it simple

Consider extensive touch-up an integral part of the paint job and avoid dark colors, faux finishes, and textures that do not touch up easily. No matter how carefully you build, it's nearly impossible to install light fixtures, countertops, plumbing, and floor finishes without dinging and damaging walls.

Whenever possible, avoid multiple colors and natural wood finishes, which complicate touch-up and add substantial labor, time, and cost. If natural woodwork is important, consider prefinished trim.

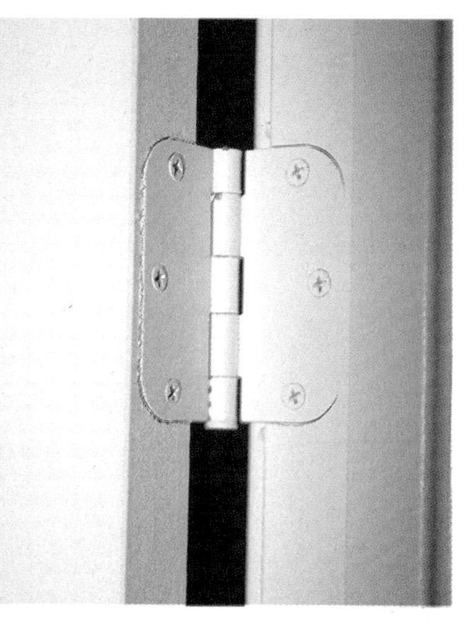

Using primed hinges saves time because doors, jambs, and casings can be painted all at the same time, eliminating at least one step.

Rolling paint is faster than masking and spraying. Darker colors may be trendy, but flat, light colors hide minor irregularities in the wall surface and make a room look larger.

TRADE SECRETS

Prefinished doors and trim made from MDF are inexpensive and renewable alternatives to tropical and hardwood millwork.

It's less expensive to trim windows with drywall than it is with wood casing.

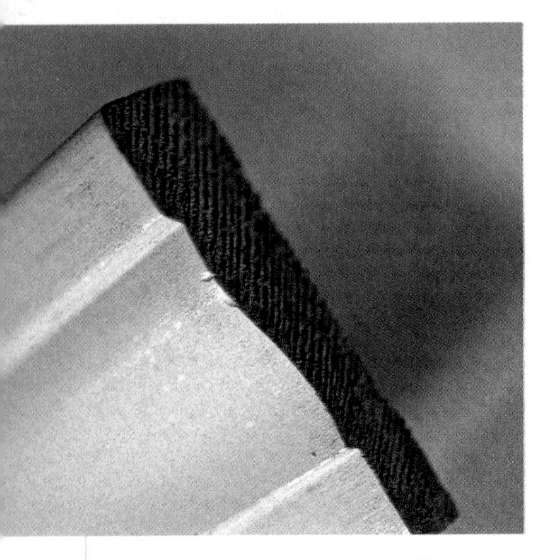

NuHome, in Houston, buys lengths of casing and uses the same material as baseboard.

Installing Interior Trim and Millwork

Molding can add interesting detail to a home's interior, but it actually serves a functional purpose by covering gaps between adjacent surfaces, such as walls and floors and windows and walls. Drywall does the same thing at a much lower cost. For example, there's no need to install wood jambs on bifold and bypass closet doors. You can omit baseboard wherever there's wall-to-wall carpeting, and you can trim windows with corner bead. You might, however, drywall-wrap three sides of a window and install a simple wood windowsill for a little variety and balance.

In general, the hierarchy of trim prices begins with MDF at the low end, and proceeds through prefinished, finger-jointed primed, poplar, and then, depending on the area, either oak or pine. When deciding whether to use MDF or a prefinished product, after you factor in the cost of painting, the new prefinished moldings become very appealing.

Economizing with casing and baseboard

If you choose to use baseboard—which, besides covering the gap between the floor and wall, also helps prevent vacuum cleaners and shoes from scuffing the paint—consider using the same trim for both base and casing. This spares material and allows you to buy one profile in greater bulk, reducing its cost.

When installing baseboard, avoid using base shoe. Nail baseboard on carpeted areas first, holding it off the floor about ½ in. On resilient and hard-surface floors, don't install the baseboard until carpenters have made their last swing through the house. This method precludes base shoe and makes the flooring installer's job a lot easier.

Prehung doors are faster

Prehung doors make light work of finishing an interior and ensure better job-site accuracy. Although hardboard slab doors cost less, many people prefer prefinished oak or wood-grained simulated panel doors. Nowadays, prefinished hardboard doors and trim look like finely stained woodwork, sans the stain, sealer, and three coats of lacquer.

Prehung doors make light work of finishing an interior and ensure better job-site accuracy.

The least expensive doors are primed hardboard. After that on the price scale come prefinished white, oak, and maple; raised-panel oak; and then luan and flat-panel birch, oak, and maple. Luan used to represent the low end in interior door skins, but no longer. It does resist scratches and dents better than hardboard, but it's also more difficult to finish. As a tropical wood, luan, which is just a cheap grade of mahogany, does not enjoy the support of environmentalists.

Low-cost closet doors

You can eliminate a door entirely on bedroom walk-in closets and just drywall-wrap the openings—closet doors remain ajar more often than not. If you decide to install a door, there's no need to install jambs. Bifold and bypass closet doors are actually designed for drywall jambs and can accommodate minor imperfections.

Sliding bypass doors are especially forgiving. Mirrored bypass doors are inexpensive, and they reflect more light into a room and enlarge its appearance. If you want to cover the track or provide a more textured, custom look, just miter-cut a piece of casing at the lintel. If you add casing on all three sides of a drywall opening and then paint it, you can achieve the appearance of a fully cased doorjamb at nearly a drywall price.

Prefinished hardboard doors and trim look exactly like finely stained woodwork but require no painting. Options include prefinished white and oak or maple look-alikes.

Using prehung doors speeds up interior finish work. These doors are being routed and prehung at Millard Lumber in Omaha, Neb.

Bedroom closet doors are often left open anyway, so eliminating them and wrapping the door opening in drywall is one way to save the cost of a door.

Until the 1970s, builders installed full-height steel sliding doors on closets. Although they cost more than hollow-core doors, you might consider the small added expense of an 8-ft. bifold versus the cost of framing a header and jambs. You can install a full-height bifold or slider wall-to-wall in a small bedroom without any closet framing at all. Steel doors come in flush, panel-embossed, and louvered finish.

> You can install a full-height bifold or slider wall-to-wall in a small bedroom without any closet framing at all.

Some builders like to use wire shelving in closets because it requires no painting. But the cost of this shelving outweighs its convenience. Simple particleboard closet shelving supported by 4-in. particleboard cleats and shelf-and-pole brackets remains the least costly way of trimming a closet. To avoid the hassle of priming and painting the shelf and pole, just use a one-coat lacquer stain to finish it. Paint the cleats along with your walls.

Window trim may be overrated

It's only in an empty house that window trim makes a striking difference. Once blinds and window treatments go up, the trim loses prominence. Drywall-wrapped window jambs and sills have become commonplace in the West and South, where aluminum and vinyl windows prevail. When builders want to dress up feature windows, such as the picture window in the living room, they install a simple wooden sill.

If your windows have wooden jambs, you can still use drywall returns instead of jamb

A length of casing can provide the textured appearance of a wood-trimmed opening without the cost of complete casing.

Mirrored closet doors are inexpensive and help to make a small bedroom look larger.

Wire shelving may seem like a bargain because it doesn't need painting or finishing, but particleboard shelves and cleats cost less.

When a stairway opens into a room, using a half-wall capped by a rail is less expensive than a conventional balustrade.

extensions by applying an L-bead along the drywall edge at the jamb and caulking the joint. Paint the jamb and bead for a three-dimensional, stepped appearance. Another way to dress up a window is to picture-frame a drywall-wrapped opening with casing. Recess the casing an inch from the edge of the drywalled jamb to create the illusion of an extra-wide molding. Paint the casing and drywall jamb with the same color and gloss for a wood-trimmed look.

You can add interior and exterior dimension to a window by wrapping a 2x4 around the outside frame, then attaching the window flange to it. The extra 1½ in. of depth lends a custom touch that comes virtually free.

You can add interior and exterior dimension to a window by wrapping a 2x4 around the outside frame, then attaching the window flange to it.

For a more dramatic effect, use a 2x4 or 2x6 on edge, but remember that larger pop-outs require a small roof over the lintel and can look a little boxy from the exterior.

Simplifying stair rails and trim

If you provide a ¾-in. drywall pad along the outside edge of the stair stringers, you don't have to install a stair skirt along the wall. When the stairway opens into a room, instead of using balusters and railing, build a half-wall, cap it with drywall or wood, and use a prefinished banister. If you prefer using a conventional railing, consider having one made of wrought iron. A standard oak banister over an iron balustrade achieves an elegant finish at reduced cost.

A bump-out for a window adds dimension to a room on both the inside and outside.

White melamine cabinets cost less than traditional cabinets made from wood.

Leaving doors off some cabinets not only saves money, but also provides display space in the kitchen.

Factory-made cabinets are less expensive than custom cabinetry and are available with high-quality, durable finishes.

Saving in the Kitchen

Kitchens have become an extension of the family room, a place to gather and entertain. Most people seem to prefer an open arrangement in which the kitchen is separated from adjacent rooms by only an island or peninsula, not a wall. The importance buyers place on a bright, spacious kitchen leads many builders to add expensive cabinetry and accessories to make their kitchens more desirable. This may appeal to some people, but many others would rather have a large built-in pantry instead of extra cabinet space, and would accept a few feet of open shelving in place of upper cupboards.

Don't be afraid to eliminate doors on a few cabinets and just provide open shelving.

Factory cabinets usually cost less than custom-made and come finished with high-quality stains and lacquer. Many people prefer a traditional appearance, but contemporary white melamine cabinets cost less. Some builders solve this problem by combining melamine boxes with wood doors. I've found that light, natural colors are the most popular.

Use stock sizes whenever possible and try to design the kitchen in 3-in. increments to correspond with fabrication standards. Try to use the widest sizes available, since wider cabinets cost less per foot than narrow ones.

Door styles influence cost more than any other element. A simple, classic design is a good place to start, but there are many upgraded patterns to choose from. Don't be afraid to eliminate doors on a few cabinets and just provide open shelving. This is especially true with corner cabinets, where it's difficult to reach stored items, and open shelving can be used for display.

Cabinets and countertops represent a significant expense. It's wise to plan the kitchen

TRADE SECRETS

To give countertops a high-end, high-gloss appearance, use a gloss gel-coat polish, such as 3M's Gloss GelCoat®.

around storage and working areas while minimizing cabinets and expensive surfaces. A pantry the size of a linen closet can replace about 6 ft. of cupboard storage space. You can add counter surface area with an inexpensive peninsula. Instead of using base cabinets, install upper cabinets over a kickboard and, if necessary, make the peninsula wider with a framed wall. Never hang upper cabinets over a peninsula, since this adds cost and clutters the view. To save money and add light, install a large window adjacent to the peninsula. This creates the illusion of space and fills a section of wall that would otherwise beg for more cupboards.

Laminate countertops are a bargain

Post-formed laminate countertops are the least expensive kitchen surface available and provide a practical decorative surface. Many people, however, prefer solid-surface countertops, such as Corian® and Surell®. Builders may offer these popular but expensive surfaces as upgrades or use the solid surface as a feature on an island or bar top.

With a routed, oak edge, this inexpensive laminate top looks more refined.

Post-formed laminate counters are the most economical option in kitchens and baths.

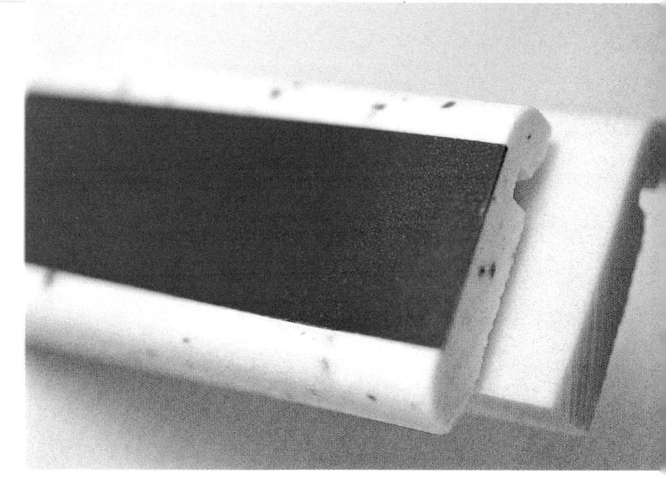

The Kuehn Bevel Co. manufactures an inexpensive edging material that can be applied to a laminate countertop to make it appear like a more costly solid surface counter.

TRADE SECRETS

Many polyester carpets are made from recycled products, such as soda bottles and milk containers. Because these fibers derive from top-quality food-grade resins, they are actually superior to the lower-grade synthetic fibers used in many polyester products. Look for the PET (polyethylene terephthalate) label on carpet label to make sure.

Only four companies produce almost all of the plastic laminates available, and their prices remain comparable. Manufacturers compete by offering new finishes rather than cheaper products. Natural stone patterns, such as granite, have become very popular. Faux granite laminate can be combined with a granite tile backsplash for a high-end look. Ceramic tile also can be used for the backsplash to dress up a laminate counter, especially if the feature surface uses the same tile. Routed wood edging or a contrasting color of laminate can add an accent that makes laminates look stylish.

Of the solid-surface options, Swanstone® offers a ⅛-in. veneered top that costs about $40 per square foot, compared to the $50 to $70 surfaces from other manufacturers.

Flooring is a key ingredient for interior finishes. Hardwood and ceramic tile are among the most expensive options; sheet vinyl and carpet are the most economical.

Flooring Makes a First Impression

The quality of the floors is apparent the moment you step over the threshold. Even more than the front door, entry foyer, or overall floor plan, flooring forms the first impression of the interior of your home. Many real-estate agents have become rich by buying inexpensive fixer-uppers, changing the floors, and selling the same house at a profit. Flooring is also the most used component of a home. Lights go on and off, appliances work when needed, but floors operate whenever occupants are home.

Perhaps more than any other finish product, flooring includes a huge array of brands, styles, and color choices. But for purposes of discussion, we can classify most flooring products into three categories: wall-to-wall carpet, resilient flooring, and hard-surface flooring. Carpet is generally preferred in the family room, bedrooms, and halls, while the kitchen and laundry get resilient flooring; hard-surface floors, such as wood and tile, go in the dining room and bathrooms, respectively. Expensive homes feature ceramic tile or hardwood in the kitchen and entry hall.

Although fewer than half a dozen companies manufacture most flooring products, intense competition within divisions and mills scattered all over the country makes for a broad array of choices. Some are extremely inexpensive. The sections that follow explain how to design for each category of flooring and what to look for when choosing inexpensive varieties.

Avoid wasting carpet with room layout

Carpet comes in 12-ft.-wide rolls. Every carpet has a weave pattern that an installer must match when creating a seam. To minimize waste and avoid having to buy excess carpet, design your rooms in 6-ft. increments and try to avoid T-shaped or L-shaped room configurations. For example, if you have a 12-ft.-wide

The polyester carpet on the right has a thicker pile and costs less than the nylon carpet on the left. But five years from now, the nylon carpet will hold its pile and still look new. The polyester carpet will mat down in high-traffic areas and appear worn out.

living area that runs into a parallel hallway, the carpet can extend right through into the hall with no waste. But when that hall leads to a 12-ft. by 8-ft. room that lies at a right angle, the installer cannot run the 12-ft. roll along the length of the room without going against the grain. Instead, the installer has to lay one 12-ft. by 8-ft. piece and then seam a second 4-ft. by 8-ft piece to complete the job. A full 8-ft. length of carpet goes to waste. You pay for this material, which your installer takes back to the showroom to sell as a "remnant."

> The quality of the floors is apparent the moment you step over the threshold.

Don't buy the most expensive carpet In most cases, it doesn't pay to buy expensive, high-durability carpet. People change carpets every 5 or 10 years because the material has become soiled, damaged, or out of style. It's rare to find carpet that's really worn out. A 5- to 10-year FHA-quality carpet generally costs between $6 and $10 per square yard and provides an attractive surface with a reasonable service life. To get a plush feel, increase the quality or thickness of the padding at minimal cost.

Although three major mills manufacture almost all of the carpet on the market, the distribution system includes stores that buy from dealers and those that buy directly from a mill. Mill-direct stores offer less selection, but at better prices. Remnants, closeouts, and sales also provide excellent buying opportunities.

Polyester represents the least expensive variety of carpet—sometimes $3 per square yard or less. Polyester can have a great "hand," as carpet dealers describe pile thickness. It has excellent stain resistance and generally comes with a 5- to 10-year warranty. But polyester does not hold up as well as nylon. The pile

TRADE SECRETS

Urethane Soy Systems Co. of Princeton, Ill., manufactures an inexpensive soy-based polymer for use in carpet backing. Look for the trademark SoyOyl™ or Dow Chemical's BIOBALANCE® on the carpet label.

Carpet Pad Counts

Residential carpets are generally installed over one of three varieties of foam pad: inexpensive foam, rebond, and fine-grind rebond, ranging in price from about $1 to $3 per square yard. Most builders choose a 7⁄16-in. rebond pad because it provides durability at a reasonable price.

Inexpensive foam pads crush easily, and carpet tends to wear out quickly when installed over it. Rebond pads, which are made from 100% recycled waste products, wear almost as well as the more expensive pads, but they can have hard objects, such as crushed bottle caps, in the material. You can feel these hard spots underfoot through certain carpets and, wherever they occur, carpets tend to wear more quickly. Fine grinds, or deluxe pads, which are made like rebond without the recycled scraps, provide a smooth, highly resilient cushion at a premium price. An inexpensive carpet on a deluxe pad can feel luxurious underfoot.

If you want carpet to feel more comfortable, you might choose a good, high-density pad, since thick, soft pads (less than 6 lb.) can reduce the lifespan of the carpet, and they can even void the warranty. Thinner pads don't sink as much with each footstep, which minimizes the pull on carpet backing, preventing the latex in the backing from breaking down as quickly. Avoid rubber pads with a waffle pattern, since these do not provide enough support. Thin, firm pads work best under wool Berber carpet and thin, commercial carpet.

From left to right: Inexpensive foam, rebond, and fine-grind rebond padding. Most builders choose 7⁄16-in. rebond pad because it costs less, wears well, and represents an environmentally responsible alternative.

tends to mat down prematurely in high-traffic areas, such as hallways. A thinner builder grade of nylon carpet lasts 5 to 7 years before showing signs of wear, while a thick polyester carpet may last half as long.

When choosing carpets, you'll see that the least expensive varieties have "stapled" fiber and the better grades have "continuous filament." The stapled variety tends to fray more quickly and releases fuzz when vacuumed. If you're looking for the most durable product at a low cost, choose the least expensive line of continuous-filament nylon available, even when it does not appear as bulky as less expensive options. Fiber can be deceiving. Longevity depends more on the fibers' length and twist than the total weight, or "plush."

Lay out resilient flooring as you would carpet

Resilient flooring comes in tile squares and rolled sheet goods. Sheet goods represent the largest market segment in residential construction. Like carpet, sheet goods have a pattern. If kitchens are designed in a 12-ft. width and bathrooms in 6-ft. lengths, it's possible to take full advantage of a seamless, one-piece floor with very little waste.

Underlayment becomes an important cost consideration whenever you're installing resilient flooring. Beyond 3⁄8-in. plywood, luan, or OSB, fiber-floor underlayment made from high-density particleboard with a waterproof coating represents the least expensive and most practical underlayment.

Cheap vinyl is no bargain Several qualities of sheet flooring are available at a range of prices. Unlike carpet, cheap sheet goods wear out quickly and often tear before construction is over. Installers have trouble getting some cheap goods into the house without ripping them.

If you choose low-end resilient flooring, the quality of the subfloor and floor prep becomes especially important. Any bumps in the floor create areas of wear, and any loose flooring becomes prone to tear. A thicker, slightly higher-quality vinyl often pays for itself.

In broad terms, you'll find two types of sheet goods: urethane and PVC. Urethane provides a more durable finish. PVC tends to scuff easily, especially with black heel marks. Some builders install urethane products in the kitchen and entry and reserve PVC for bathrooms, laundry rooms, and areas subject to less wear.

> Unlike carpet, cheap sheet goods wear out quickly and often tear before construction is over.

Tile and hardwood are the most costly options

The most common application of hard flooring is in the entry foyer, where even the least expensive homes often feature ceramic tile. In warmer climates, ceramic tile is the preferred surface for entries, kitchens, and bathrooms. In colder regions, buyers choose hardwood for the same areas. Both materials cost more than resilient flooring.

You can provide the illusion of hardwood flooring by installing it around the perimeter of a room and then carpeting the center. This works especially well in a dining room, where the carpet looks like a throw rug used to outline and accent the dining-room table. Consider lower-grade oak flooring, too, as slight imperfections in the wood are often preferable to no hardwood flooring at all.

The least expensive hardwood alternative is plastic laminate. This newcomer to the flooring industry provides a long-lasting,

Apartment-quality sheet goods, such as Armstrong's Apartment® 1, shown here, tear easily. At a minimum, choose a builder's line, such as Armstrong's Initiator® or Mannington's Vega II® series.

TRADE SECRETS

At a minimum, choose a 10-mil builder-grade PVC resilient flooring, such as Mannington's Vega II, and avoid "apartment quality." Although builder-grade vinyl costs $2 or $3 more per square yard, it's a better buy.

Hardwood flooring, like the oak in this hallway, is a relatively expensive flooring option. But it can be used selectively, and a lower grade of flooring may be perfectly adequate.

Plastic laminated flooring products, like these made by Formica, provide an excellent low-cost alternative to veneered and solid hardwood floors.

Floor tiles include many options, from solid limestone to sun-baked mud pavers. Glazed ceramic tile is the most popular. Among the residential ceramics, clay-back tiles, such as the one shown on the left, cost less than the "through-back" solid-color tile on the right.

good-looking floor product that installs so easily that many homeowners put it in by themselves. If you decide to use this product in a wet area, such as the kitchen or laundry room, make sure to choose a brand with a laminated, high-density backing. Some of the less expensive varieties have particleboard backing, which can swell when wet. This can cause problems not only in wet areas, but also over damp concrete slabs. Generally, avoid these laminated flooring products in bathrooms, because even with high-quality backing, the panel ends can take on water and swell.

> Throughout most of the United Sates, ceramic tile floors are four times as expensive as vinyl.

Other flooring alternatives include environmentally friendly products such as bamboo and cork, which certainly cost more than vinyl or plastic laminates, but a little less than ceramic tile and veneered hardwoods. In some areas, solid bamboo planks cost less than oak. As renewable products become more popular, their cost should drop and become highly competitive.

Labor costs a variable with ceramic tile The cost of ceramic tile depends largely on labor, so it varies from region to region, depending on the cost of living and the number of skilled tradespeople competing for work. Throughout most of the United States, ceramic tile floors are four times as expensive as vinyl. This is especially true where crawl-space and basement homes require a concrete-board underlayment. Over slabs, ceramic tile provides a more economical surface.

Tile Look, Concrete Price

In Dallas, affordable homebuilder Carl Franklin Homes experimented successfully with floor slabs saw-cut in a grid pattern, stained, and polished to create tile-like surfaces for kitchens, bathrooms, and entryways at vinyl-floor prices. Painted or stained concrete also provides an excellent basement or laundry-room floor. Use penetrating stains designed for concrete because they don't scuff and wear off quickly. It's not necessary to use the high-tech epoxy products developed for industrial applications—in spite of what the paint dealer tells you. After the stain dries, wax the concrete for a lustrous, durable finish.

Appendix 1:
The Efficiency of Shapes

These five commonly-shaped houses illustrate how the basic blueprint has a direct impact on costs, even before you start refining your cost-saving strategies.

1. Square house

Size	Unit cost	Total	Comment
1,800-sq.-ft. foundation	$8.79	$15,822	
1,800-sq.-ft. roof	$8.24	$14,832	20% more expensive than shapes listed below due to wide span
170 ft. of wall	$38.00	$6,460	
4 corners	$55.68	$223	
0 valleys	$73.78		
2 gable ends	$208.00	$416	
Total shell cost:		$37,753	When square-shaped homes exceed 32 ft., they become slightly more expensive to build.

2. Rectangular house

Size	Unit cost	Total	Comment
1,800-sq.-ft. foundation	$8.79	$15,822	
1,800-sq.-ft. roof	$6.87	$12,366	
180 ft. of wall	$38.00	$6,840	
4 corners	$55.68	$223	
0 valleys	$73.78		
2 gable ends	$208.00	$416	
Total shell cost:		$35,667	

3. L–shaped house

Size	Unit cost	Total	Comment
1,800-sq.-ft. foundation	$8.79	$15,822	
1,800-sq.-ft. roof	$6.87	$12,366	
220 ft. of wall	$38.00	$8,360	
6 corners	$55.68	$334	
2 valleys	$73.78	$148	
3 gable ends	$208.00	$624	
Total shell cost:		$37,654	

4. C–shaped house

Size	Unit cost	Total	Comment
1,800-sq.-ft. foundation	$8.79	$15,822	
1,800-sq.-ft. roof	$6.87	$12,366	
228 ft. of wall	$38.00	$8,664	
8 corners	$55.68	$445	
4 valleys	$73.78	$295	
4 gable ends	$208.00	$832	
Total shell cost:		$38,424	

5. Two-story square house

Size	Unit cost	Total	Comment
900-sq.-ft. foundation	$8.79	$7,911	
900-sq.-ft. roof	$6.87	$6,183	When you build two stories, you can build larger without overspan.
240 ft. of wall	$38.00	$9,120	
8 corners	$55.68	$445	Per 8 ft. wall
0 valleys	$73.78		
2 gable ends	$208.00	$416	
Total shell cost:		$24,075	Two-story houses have more wall area, but they make up for this with half the foundation and floor.

Unit costs derived from *Contractor's Pricing Guide: Residential Square Foot Costs 2001* by Robert S. Means (Robert S. Means Company, 2001.)

Appendix 2:

Example of Insulation Schedule and Suggested HVAC Sizing for 1,680-sq.-ft. Four-Bedroom Home in Climate Zone 14.

Each cardinal direction has slightly different sizing requirements. In this instance, east and west exposures have higher cooling loads. Changes in the R-values of the insulation and windows can affect the size of your HVAC requirements dramatically. In this instance, low-e glazing reduces the heating load by about 10,000 lb. This translates into a smaller furnace. A good insulation and infiltration package reduces the cooling load by as much as a ton. You'll see that the optimum cost-to-value package in this instance (highlighted) is not the highest-value insulation package, but the least expensive insulation and window combination that achieves the goal of reducing HVAC requirements to a minimum.

Area	Type	Standard Material	#1 Optional Material	
Exterior walls (main level)	Batt	R-13	R-15	$75.00
Ceiling (flat)	Blown-in	R-30	R-45	$65.00
Ceiling (vaulted)	Batt	R-30	R-45	$100.00
Box sill	Batt	R-11	R-19	$35.00
Windows		Double-glazed	Low-e	$100.00
Infiltration				
Front of home	**Direction**	**South**	**South**	
Heating load		32,347	24,163	
Furnace size		15 kw	10 kw	($250.00)
Cooling load		19,728	17,081	
Heat pump size		2 ton	1.5 ton	($250.00)
				($125.00) savings
Windows		Double-glazed	Low-e	
Front of home	**Direction**	**South**	**South**	
Heating load		32,347	24,163	
Furnace size		15 kw	10 kw	($250.00)
Cooling load		19,159	16,586	
Heat pump size		2 ton	1.5 ton	($250.00)
				($125.00) savings
Windows		Double-glazed	Low-e	
Front of home	**Direction**	**East or west**	**East or west**	
Heating load		32,347	24,163	
Furnace size		15 kw	10 kw	($250.00)
Cooling load		24,180	21,239	
Heat pump size		2.5 ton	2 ton	($250.00)
				($125.00) savings

Suggested HVAC sizing is abased on Manual J load calculations.

Appendix 3:
Joist-Span Table

Sample floor-joist spans for major species based on a live load of 40 lb. per square foot and a dead load of 10 lb. per square foot. Find the least expensive joist that bridges the width required.

Size	Spacing	Grade	Hem/fir	Spruce/pine/fir	Douglas fir & larch	Southern yellow pine
2x6	16 in.	#1	9 ft. 6 in.	9 ft. 4 in	9 ft. 11 in.	9 ft. 11 in
		#2	9 ft. 1 in.	9 ft. 4 in.	9 ft. 9 in.	9 ft. 9 in.
	24 in.	#1	8 ft. 4 in.	8 ft. 1 in.	8 ft. 8 in.	8 ft. 8 in.
		#2	7 ft. 11 in.	8 ft. 1 in.	8 ft. 1 in.	8 ft. 6 in.
2x8	16 in.	#1	12 ft. 7 in.	12 ft. 4 in.	13 ft. 1 in.	13 ft. 1 in.
		#2	12 ft. 0 in.	12 ft. 4 in.	12 ft. 7 in.	12 ft. 10 in.
	24 in.	#1	10 ft. 9 in.	10 ft. 3 in.	11 ft. 0 in.	11 ft. 5 in.
		#2	10 ft. 2 in.	10 ft. 3 in.	10 ft. 3 in.	11 ft. 0 in.
2x10	16 in.	#1	16 ft. 0 in.	15 ft. 5 in.	16 ft. 5 in.	16 ft. 9 in.
		#2	15 ft. 2 in.	15 ft. 5 in.	15 ft. 5 in.	16 ft. 1 in.
	24 in.	#1	13 ft. 1 in.	12 ft. 7 in.	13 ft. 5 in.	14 ft. 7 in.
		#2	12 ft. 5 in.	12 ft. 7 in.	12 ft. 7 in.	13 ft. 2 in.

Index

Q

R

S

T

U

V

W